Yuri Bozzi

A study on the activity-dependent expression of neurotrophic factors in the rat visual system

TESI DI PERFEZIONAMENTO

SCUOLA NORMALE SUPERIORE
1997

Tesi di perfezionamento in Neurobiologia sostenuta il 25 marzo 1996

COMMISSIONE GIUDICATRICE

Lamberto Maffei, Presidente
Giuseppina Barsacchi
Nicoletta Berardi
Maria Luisa Carrozza
Laura Colombaioni
Federico Cremisi
Luciano Domenici
Dan Lindholm

ISBN: 978-88-7642-272-0

A Simona

TABLE OF CONTENTS

SUMMARY

The mammalian visual system is a classical model for the study of the mechanisms that regulate the development and refinement of connections in the Central Nervous System (CNS). It is well known, from a large series of studies, that visual deprivations obtained by closing one eye (monocular deprivation) or by rearing animals in complete darkness from birth can irreversibly impair both physiological and anatomical properties of the visual cortex. Visual deprivations are particularly effective when they are performed in young animals. Therefore, a correct visual experience during a short period of early postnatal life (called "critical period") is necessary for the functional and anatomical development of the mammalian visual system. This is a general feature of CNS: sensory experience can dinamically modify both the anatomy and function of the brain. In particular, a large series of studies over the past thirty years has largely demonstrated that neurons can modify their connections in response to environmental stimuli. This process, generally referred to as "neuronal" (or "synaptic") "plasticity", is the basis of learning and memory processes in the brain.

Which are the molecular bases of neuronal plasticity? That is, how can environmental factors and sensory experience modify the connections between neurons? In the brain, inputs from the surrounding environment act by modulating the basal levels of neuronal electrical activity. Many classical studies have shown that a correct electrical activity is required for the formation and stabilization of the connections between neurons. Therefore, our knowledge of neuronal plasticity is essentially based on the understanding of the cellular mechanisms that allow neuronal electrical activity to modify synaptic connections.

Neurotrophic factors of the Nerve Growth Factor (NGF) family are known to finely regulate the development and maintenance of synaptic connections in the Peripheral Nervous System. A large series of recent studies has also demonstrated that these molecules could regulate the development and plasticity of specific neuronal populations in the CNS. One of the first demonstrations for such a role came some years ago from the group of L. Maffei. The results obtained by these Authors firstly suggested the hypothesis that neurotrophic factors could play a key role in the plasticity of the mammalian visual system. In two important studies,

Maffei and coworkers demonstrated that exogenous administration of large doses of NGF prevented the physiological and anatomical effects of monocular deprivation in rats (Domenici *et al.*, 1991; Maffei *et al.*, 1992). To explain these effects, the Authors proposed a "neurotrophic hypothesis" for the plasticity of the visual system (Maffei *et al.*, 1992). Following this theory, the formation and stabilization of geniculo-cortical connections would be regulated by a neurotrophic factor produced in limited amount by target cortical cells. The production and/or release of this factor, and possibly its uptake, could be dependent upon afferent electrical activity that is known to influence the functional development of the visual system.

At that time, the distribution of neurotrophic factors and their receptors in the visual system was not known. Some data from our and other laboratories suggested that NGF was present in the visual cortex, but a clear demonstration was lacking for the presence of an active NGF receptor molecule in the visual system. Few data were also available for activity-dependent regulation of neurotrophin expression in the CNS. These data principally came from *in vitro* studies on neuronal cultures or *in vivo* studies that used pharmacological treatments to increase neuronal activity in the brain. Finally, there was no evidence that physiological neuronal activity could control neurotrophin production in the CNS.

In our laboratory, we therefore decided to investigate whether neurotrophins and their receptors could be physiologically regulated by sensory experience in the visual system. To test this hypothesis, we used the classical experimental protocol of monocular deprivation (MD) in the rat. MD has been largely used as a model to study activity-dependent plasticity of the visual system, and its effects have been ascribed to a reduced production of an endogenous neurotrophic factor of the NGF family (Maffei *et al.*, 1992). In my study, I analysed the distribution of Brain-derived Neurotrophic Factor (BDNF) and its receptor TrkB in the rat visual system and I demonstrated that monocular deprivation is sufficient to reduce the expression of BDNF mRNA in the rat visual cortex (Bozzi et al., 1995). These results clearly show that sensory experience can regulate the production of neurotrophic factors in the visual system.

The purpose of this Thesis is to give a general overview about the recent advances in the field of neurotrophin-regulated plasticity of the mammalian visual system. During the past five years, several other Authors have been studying the distribution, function and mechanism of action of neurotrophic factors in the mammalian visual system. Taken together, these results strongly confirmed the "neurotrophic hypothesis" for the plasticity of the mammalian visual system originally proposed by Maffei and coworkers (Maffei *et al.*, 1992). All these results are reported in Chapter 1 (Introduction), where I describe the general features of the expression and mechanism of action of neurotrophic factors and their receptors in the CNS. Chapter 2 (Results) summarizes the studies that we performed in our laboratory during the past five years, in order to understand the physiological mechanisms that regulate neurotrophin expression in the rat visual system. Some of these results have been already published elsewhere (Bozzi *et al.*, 1995). In Chapter 3 (Discussion) I will present the most recent hypotheses that have been proposed for the role of neurotrophic factors in the development and plasticity of the mammalian visual system.

1. INTRODUCTION

1.1   Neurotrophic factors and plasticity of the visual system.

We generally refer to "synaptic plasticity" as the fine process through which neurons can modify their synaptic connections in response to appropriate environmental stimuli. In the Central Nervous System (CNS), inputs from the surrounding environment essentially act by modulating the basal level of neuronal electrical activity. A major postulate in the study of neuronal plasticity is that electrical activity is necessary for a correct formation and consolidation of neuronal connections.

The mammalian visual system is a classical model for the study of activity-dependent mechanisms that regulate the development and refinement of connections in the CNS. Many studies over the past 35 years have demonstrated that deprivation of visual experience during a "critical period" of early postnatal life can dramatically affect both the connectivity and perceptive properties of visual cortical neurons in mammals. In more recent years, neurotrophic factors belonging to the family of Nerve Growth Factor (NGF) have been considered as possible mediators of activity-dependent neuronal plasticity. In this Introduction I will review the classical studies on the plasticity of mammalian visual system, and I will also discuss the recent findings that suggest a possible role of neurotrophic factors in this process.

1.1.1   The visual system: a suitable model to study neural plasticity.

In the mammalian CNS, the major structures involved in visual perception are the retina, the dorsal lateral geniculate nucleus (dLGN) in the thalamus and the primary visual cortex (area 17). Axons from retinal ganglion cells form the optic nerve fibres, which cross at the level of the optic chiasm. In higher mammals (cat, monkey and man) this crossing is not complete. Crossed and uncrossed fibres from both eyes make their synaptic connections with geniculate neurons on the contralateral and ipsilateral side of the thalamus, respectively. So, each dLGN is functionally connected with both eyes. However, within the dLGN, axons from the two eyes terminate in alternating eye-specific layers that are strictly monocular. Therefore, inputs from the two eyes are segregated at the level of dLGN, and so they remain until they reach the visual cortex. In layer IV of the primary visual cortex of many mammals (such as cats and monkeys), geniculate axon terminals are segregated into alternating and eye-specific regions ("ocular dominance columns"). Figure 1.1A shows a simplified diagram of the mammalian visual system.

The formation of connections in the visual system is a fine process that takes place during the late embryonic and early postnatal development in mammals. Remarkably, neither eye-specific layers within the dLGN nor ocular dominance columns in the cortex are present initially during development (for a review see Shatz, 1990, and references therein). When retinal ganglion cell axons from the two eyes first reach their targets into the dLGN, they are intermixed with each other. The characteristic eye-specific laminar structure of dLGN slowly forms

---

Figure 1.1.

Schematic diagram of the developing mammalian visual system.

a) Anatomical organisation of the visual pathway. Fibres from the two eyes are indicated. dLGN, dorsal lateral geniculate nucleus. b) Segregation of ocular dominance columns in the primary visual cortex slowly occurs during postnatal development and depends on visual experience. In newborn mammals, primary visual cortex is not yet subdivided into eye-specific columns. Axon terminals of geniculo-cortical fibres driven by the two eyes are still intermixed within each other in the layer IV of the primary visual cortex ("newborn"). During a "critical period" of early postnatal life, the visual cortex acquires its definitive anatomical structure and becomes subdivided into "ocular dominance columns" ("adult"). This process is supposed to require a dynamic synapse reorganisation of geniculo-cortical connections and is based on activity-dependent competitive interactions between the two eyes. If the animal is deprived of vision in one eye by closing the eyelids at birth for several days to weeks ("monocular deprivation"), the cortical territories occupied by dLGN axons functionally connected with the closed eye are strongly reduced. Conversely, rearing animals in complete darkness from birth ("dark rearing") does not allow ocular dominance columns to form. See text for further details. (Modified from Shatz, 1990).

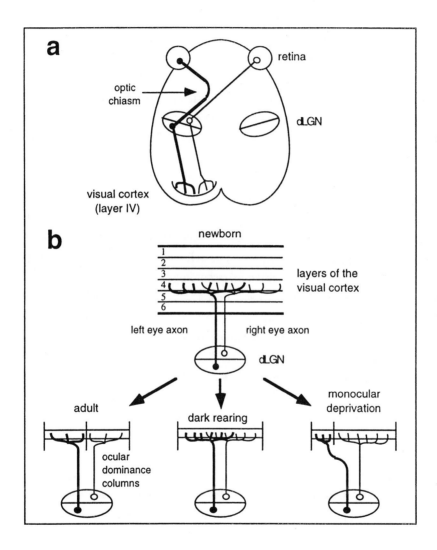

Figure 1.1.

Schematic diagram of the developing mammalian visual system.

during prenatal development. This process is supposed to require a dynamic synapse reorganization of ganglion cells terminals (Sretavan and Shatz, 1986). Ocular dominance columns formation in layer IV of primary visual cortex probably requires a similar mechanism of axonal remodelling and synapse elimination. In cats, this process is almost entirely postnatal, whereas in monkeys it occurs during the late embryonic and early postnatal life (Rakic, 1976). This was originally demonstrated by LeVay et al. (1978, 1980), by using a transneuronal radioactive tracer (tritiated proline) injected in one eye at birth. Autoradiographic exposures of visual cortex slices from cats and monkeys at different postnatal ages revealed that the segregation of geniculo-cortical fibres into eye-specific colums slowly occurs during postnatal development.

Specific functional properties of visual cortical neurons correspond to this peculiar anatomical organization of primary visual cortex in mammals. In their pioneering studies during the early 1960's, David Hubel and Thornsten Wiesel first demonstrated that mammalian visual cortex has a columnar organization (Hubel and Wiesel, 1963). By the means of extracellular recordings from the cat occipital cortex (area 17), these Authors showed that cells in primary visual cortex were organized into distinct groups (columns) according to their ability to respond to simple visual stimuli (luminous bars) that had a specific spatial orientation. Moreover, Hubel and Wiesel could also classify visual cortical neurons into seven distinct "ocular dominance classes" according to their response to visual stimulation of one eye or the other. Cells in class 1 could exclusively respond to the contralateral eye, whereas neurons in class 7 were exclusively responsive to visual stimuli presented to the ipsilateral eye. Cells in class 4 were strictly binocular, in that they could equally respond to both eyes. Finally, cells in ocular dominance classes 2-3 and 5-6 were mainly dominated by the contralateral and ipsilateral eye, respectively. Interestingly, Hubel and Wiesel found that neurons in layer IV of primary visual cortex were strictly monocular, and were organized into eye-specific columns ("ocular dominance columns"). Conversely, neurons in superficial (II-III) or inner (V-VI) layers responded to visual stimulation of either eye (Hubel and Wiesel, 1963).

Which are the mechanisms that regulate the functional and anatomical development of the primary visual cortex? How do dLGN axons representing the two eyes segregate into eye-specific columns? A first answer to these questions came from the studies of Hubel and Wiesel on the effects of visual deprivation in kittens. As these Authors demonstrated in their first studies, the majority of neurons in the normal adult visual cortex are binocular: that is, they respond to visual stimulation of either eye. If the animal is deprived of vision in one eye by closing the eyelids at birth for several days to weeks ("monocular deprivation", MD), the ocular dominance distribution of neurons in visual cortex is dramatically shifted. The percentage of cortical cells that respond to deprived eye is strongly reduced, and the great majority (90%) of neurons are monocularly driven by the stimulation of the open eye (Wiesel and Hubel, 1963b). The same Authors also demostrated that neurons in the dLGN connected to the deprived eye have a strong reduction in their soma size (25-40% of "shrinkage"), whereas their response to normal stimuli remains unaltered (Wiesel and Hubel, 1963a).

The physiological shift in ocular dominance classes is paralleled by a profound change in the anatomical organization of dLGN axons within layer IV of the visual cortex. Monocular deprivation results in a dramatic reduction of the cortical territories (ocular dominance columns) occupied by dLGN axons functionally connected with the closed eye, and also causes an expansion of the neural territory innervated by the open eye (Fig. 1.1B). This was clearly demonstrated by anterograde labelling experiments with tritiated proline injected in the open eye of MD cats and monkeys (Hubel *et al.*, 1977; Shatz and Stryker, 1978; LeVay *et al.*, 1980).

In their initial paper, Wiesel and Hubel also reported that monocular deprivation did not cause any anatomical or physiological deficit in adult cats (Wiesel and Hubel, 1963b) (Figure 1.2). Therefore, there is a "critical period", early in postnatal life, when the normal development and maintenance of connections in the mammalian visual system is susceptible to dramatic alterations by abnormal visual experience (see also Hubel and Wiesel, 1970). But how might abnormal visual experience (such as eye closure) influence the connectivity within the visual cortex? Many studies support the hypothesis that the formation of ocular dominance columns during the critical period is regulated by an use-dependent synaptic competition between afferent fibres from the two eyes onto layer IV cortical neurons. In the cat, binocular eye closure from birth (binocular deprivation, BD) does not affect the distribution of ocular dominance classes, that remains unaltered as compared to that observed in newborn kittens (Wiesel and Hubel, 1965). Therefore, the simultaneous closure of both eyes, that abolishes competition between the two afferent inputs, does not allow segregation to take place. The same results were also obtained by Stryker and Harris (1986), who demonstrated that when inputs from both eyes are completely silenced by injecting tetrodotoxin (TTX, a sodium channel blocker) for several weeks, ocular dominance columns do not form, and the great majority of cortical cells still remain binocular, as in newborn animals. Finally, if artificial strabismus is produced by cutting the extraocular muscles of one eye during the critical period, binocular vision is completely impaired, and every cortical neuron becomes exclusively monocularly innervated (that is, all cortical cells can be essentially grouped in the two "monocular" classes 1 and 7; Hubel and Wiesel, 1965) (Figure 1.2).

Therefore, we can generally conclude that neural activity is necessary for the correct formation and consolidation of geniculo-cortical connections. The timing and patterning of retinal neuronal activity is also crucial for a correct segregation of ocular dominance columns, as demonstrated by an experiment by Stryker and Strickland (1984). These Authors used binocular injections of TTX to completely block afferent activity to the visual cortex, and then artificially stimulated optic nerve fibres with extracellular electrodes. Synchronous stimulation of the two nerves prevented segregation of dominance columns, whereas asynchronous stimulation allowed them to form. In subsequent studies, Reiter and Stryker (1988) also demonstrated that the coincident activation of presynaptic and postsynaptic elements in the visual cortex is necessary for a correct segregation. When cortical neurons are silenced by intracortical infusion of muscimol (an agonist of γ-aminobutyric acid type A -$GABA_A$- receptors), monocular deprivation results in a shift of ocular dominance in favor of the *closed* eye (Reiter and Stryker, 1988). These results strongly suggest that in the visual cortex, the concurrent activation of both pre- and post-synaptic neurons can strengthen their synaptic connections, whereas synapses are weakened by lack of coincident activation. These observations are reminiscent of a Hebbian rule for synapse consolidation (Hebb, 1949). However, it is still largely unknown which are the cellular and molecular mechanisms that regulate synapse strengthening in the visual cortex. In the recent years, several studies have demonstrated that the N-methyl-D-aspartate (NMDA) glutamate receptor subtype could be involved in the process of dominance column segregation. For example, when NMDA receptors located on visual cortical neurons are blocked by intracortical infusion of the NMDA antagonist 2-amino-5-phosphonovaleric acid (APV), monocular deprivation causes an ocular dominance shift in favor of the *closed* eye (Bear *et al.*, 1990), as already observed by Reiter

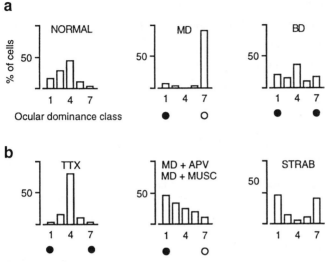

Figure 1.2.

The physiological properties of visual cortical neurons depend on afferent activity. Electrophysiological recordings from the cat and monkey visual cortex have been used by several Authors to assess the effect of different manipulations on the ocular dominance of visual cortical neurons. Ocular dominance classes (ODC) are here indicated according to the classification by Hubel and Wiesel. Cells in classes 1 and 7 respond to visual stimulation of the contralateral and ipsilateral eye, respectively. Cells in ODC 4 can equally respond to both eyes. a) In the adult visual cortex of cats and monkeys, the majority of neurons are binocular (NORMAL). Closing one eye during the early postnatal life (monocular deprivation, MD) markedly shifts this distribution in favor of the open eye (open circle). Conversely, binocular eye closure (binocular deprivation, BD) essentially leaves ODC unaltered. b) When inputs from both eyes are completely silenced by intraocular injections of tetrodotoxin (TTX), segregation does not occur, and the great majority of cortical cells still remain binocular, as in newborn animals. Coincident activation of presynaptic and postsynaptic elements in the visual cortex is necessary for a correct segregation. When cortical neurons are silenced by intracortical infusion of muscimol or APV, monocular deprivation results in a shift of ocular dominance in favor of the *closed* eye (MD+MUSC, MD+APV). Artificial strabismus (STRAB) completely impairs binocular vision and every cortical neuron becomes exclusively monocularly innervated. (Modified from Shatz, 1990).

and Stryker (1988) with muscimol infusion. This result strongly confirmed the idea the concurrent activation of both pre- and post-synaptic neurons is necessary for the strengthening of synaptic connections, and also provided the first clue for a direct role of NMDA receptors in visual cortical plasticity. The results of these experiments are summarized in Fig. 1.2.

The requirement of NMDA receptor activation for a correct segregation of geniculo-cortical connections has been also suggested by several studies on the induction of long-term potentiation (LTP)* in the visual cortex. Taken together, all these data support the hypothesis that multiple phenomena concurr in the fine process of input segregation that takes place in the mammalian visual cortex during the critical period.

1) Neural afferent activity is necessary for a correct segregation. In particular, the timing and patterning of afferent inputs seems to be important.

2) The coincident activation of both pre- and post-synaptic neurons in the visual cortex also plays a critical role. NMDA receptors could be the molecular

---

* Long-term potentiation (LTP) (reviewed in Bliss and Collingridge, 1993) is the best understood form of activity-dependent synaptic plasticity. LTP was first described in 1973 by Bliss and Lomo, who observed that brief trains of high-frequency stimulation to monosynaptic pathways in the hippocampus result in a rapid and long-lasting increase of synaptic transmission (Bliss and Lomo, 1973). Induction of LTP normally requires synchronous presynaptic activity and postsynaptic depolarization, both concurring to activate NMDA glutamate receptors channels. Presynaptic activity is necesssary to induce the release of glutamate and its binding to postsynaptic NMDA receptors, while postsynaptic depolarization is required to release the voltage-dependent block of NMDA channels by Mg^{2+}. These two events allow Ca^{2+} ions to pass through the channels, thus triggering the biochemical pathways that lead to enhanced activity of the active synapses. One of the proposed mechanisms involves the Ca^{2+}/ calmodulin - induced activation of nitric oxide synthase (NOS) in the postsynaptic cell. NOS thus produces nitric oxide (NO), that can retrogradely diffuse from postsynaptic to presynaptic terminals, serving to stabilize them (reviewed in Schuman and Madison, 1994). A large series of studies have demonstrated that LTP can be induced in visual cortical slices (reviewed in Shatz, 1990; Fox and Zahs, 1994; Cramer and Sur, 1995; Fox, 1995).

"tranducers" required for synapse strengthening.

3) Finally, a physiological competition between inputs from the two eyes is required for ocular segregation, as suggested by the observation that abnormal competitive interactions (such those produced by MD or strabismus) dramatically alter ocular dominance columns. Therefore, what are cortical synapses competing for? Is there a competition for a diffusible factor produced by cortical neurons? A large series of studies during the past recent years strongly support this hypothesis.

1.1.2   The rat visual system.

In the recent years, also the rat visual system has been extensively studied as a model for activity-dependent neuronal plasticity. The anatomy of the rat visual system differs from that of higher mammals. Nevertheless, many of its physiological properties are very similar to those observed in the visual system of cats and monkeys. In the rat, left and right visual cortices receive afferents from both eyes, but the great majority (95-98%) of fibres from retinal ganglion cells crosses at the level of the optic chiasm (Sefton and Dreher, 1985). Moreover, the dorsal lateral geniculate nucleus (dLGN) does not present the classical laminar structure, nor ocular dominance columns have been observed in the visual cortex. In the dLGN, two different areas can be identified: the outer ("contralateral lamina"), that receives afferent fibres from the contralateral eye, and the inner ("ipsilateral lamina"), which is connected to the ipsilateral one (Reese, 1988). All cells in the ipsilateral lamina project exclusively to the binocular portion of the visual cortex (see below), whereas the contralateral lamina is connected to both monocular and binocular subfields. However, the areas in the contralateral lamina that project to the two regions of the visual cortex are anatomically distinct. Therefore, also in the rat, inputs from the two eyes are segregated at the level of dLGN, and so they remain until they reach the visual cortex. In the primary visual cortex, two different areas have been identified by using a transneuronal radioactive tracer (tritiated proline) injected into the eye (Zilles et al., 1984): the monocular (Oc1M, medial) and binocular (Oc1B, lateral) subfields, which receive afferent fibres from one or both eyes, respectively. No alternate patches, functionally connected to one eye or the other (ocular dominance columns), have been identified.

The physiological properties of neurons in rat primary visual cortex show many similarities with those observed in cats and monkeys. According to the Hubel and Wiesel classification, rat cortical neurons have been subdivided in seven different ocular dominance classes (Maffei *et al.*, 1992). Cells in ocular dominance class 1 and 7 are exclusively responsive to the contralateral and ipsilateral eye, respectively. Cells in ocular dominance class 4 can equally respond to both eyes. Finally, cells in ocular dominance classes 2-3 are mainly dominated by the contralateral eye and cells in classes 5-6 are mainly dominated by the ipsilateral eye. In the adult rat, the ocular dominance distribution shows a clear dominance of the input from the contralateral eye: ~50% of the cells are in class 2-3 and ~20% in class 1. This distribution is consistent with the observation that the great majority of retinal axons crosses at the level of the optic chiasm. In spite with the small number of uncrossed fibres, the percentage of binocular cells (classes 2-6) is quite high (~80% of the total distribution; Maffei *et al.*, 1992) and is comparable to that observed in cats and monkeys. This observation does not completely reflect the anatomical organization of the rat primary visual cortex: in fact, Zilles *et al.* (1984) observed that the monocular subfield is apparently larger than the binocular one.

As in other mammals, the responsive properties of rat visual cortical neurons gradually appear during the first postnatal weeks. At postnatal day (P) 19, the great majority of cells (~90%) have no orientation preference and are binocularly driven. At P27, the percentage of nonorientational and binocular cells is markedly reduced to 35% and 80%, respectively. The percentage of monocular cell shifts from 46% (P19) to 65% (P27) (Maffei *et al.*, 1992). Values observed in adult rats (> P45) do not significantly differ from those observed at P27. Therefore, the most important physiological properties of rat visual cortical neurons rapidly mature between the second and fourth postnatal week, and then remain essentially unchanged.

1.1.3   The effects of monocular deprivation in the rat visual system.

As in higher mammals, monocular deprivation during the first postnatal weeks dramatically affects the physiology and anatomy of the rat visual system. The effects of monocular deprivation in the rat have been extensively studied by Maffei and coworkers in the recent years (see Maffei *et al.*, 1992, for the visual cortex; see also Domenici *et al.*, 1993 for a review of the data on dLGN).

In a first series of experiments, these Authors demonstrated that monocular deprivation during the first postnatal weeks has marked amblyopic effects: visual acuity of the deprived eye estimated by visual evoked potentials

(VEPs) is strongly reduced (Domenici *et al.*, 1991). In their following paper, Maffei and coworkers extended their observations to the visual cortex. In the visual cortex contralateral to the closed eye, monocular deprivation results in a marked shift of ocular dominanace distribution (Maffei *et al.*, 1992). The contralateral, deprived eye drives the response of only 16% of cortical neurons (classes 1-3), and the ipsilateral eye dominates the great majority of cells (75% in classes 5-7). The percentage of binocular cells is reduced from 80% to 40%.

The effects of monocular deprivation are not restricted to the visual cortex. In dLGN, monocular deprivation during the first postnatal weeks causes a significant reduction in the cell soma size ("shrinkage") of the neurons in the laminae connected to the deprived eye (Domenici *et al.*, 1993).

As in other mammals, also in the rat visual cortex exists a sensitive (critical) period for monocular deprivation. This critical period spans the third, fourth and fifth postnatal weeks (P15-35) (Fagiolini *et al.*, 1994a). Monocular deprivations starting after P45 have no effect on ocular dominance distribution of visual cortical neurons (Domenici *et al.*, 1994a).

1.1.4   The effects of Nerve Growth Factor in monocularly deprived rats and the "neurotrophic hypothesis" for the plasticity of the visual system.

In a large series of recent studies, Maffei and coworkers have provided strong evidence that exogenous administration of large doses of Nerve Growth Factor (NGF) prevent the physiological and anatomical effects of monocular deprivation in rats and cats. In their first study, they demonstrated that the visual acuity of monocularly deprived rats treated with NGF was comparable to that of normal, nondeprived, rats (Domenici *et al.*, 1991). In the visual cortex, NGF prevented the shift of ocular dominance distribution caused by monocular deprivation (Maffei *et al.*, 1992). These results are summarized in Fig. 1.3. In the rat dLGN, NGF prevented the shrinkage of neurons in the laminae connected to the deprived eye (Domenici *et al.*, 1993). In all these experiments, rats were monocularly deprived from postnatal day 14 (immediately before eye-opening) to P45, and injected into the lateral ventricle with NGF. NGF treatment lasted the whole deprivation period. Further experiments confirmed these findings also in kittens (Carmignoto *et al.*, 1993a; Fiorentini *et al.*, 1995).

To explain the effects of NGF on monocularly deprived animals, Maffei and coworkers proposed a "neurotrophic hypothesis" for the plasticity of the visual system (Maffei *et al.*, 1992). They postulated that during the formation of geniculo-cortical connections, afferent fibres from the dLGN could compete for a neurotrophic factor produced in limited amount by target cortical cells. The production and release of this factor, and possibly its uptake, could be dependent upon afferent electrical activity. Following this theory, electrical activity in the fibres driven by the deprived eye would be insufficient or inappropriate for the necessary production and/or uptake of the neurotrophin. Hence, deprived fibres without enough neurotrophic factor would loose the competition with the non deprived eye unless neurotrophic factor is provided exogenously.

NGF belongs to a large family of trophic factors, also known as neurotrophins, that share many structural and functional properties and can cross-react, at least *in vitro*, with the same receptor molecules (see below, in this Introduction). Therefore, considering the high doses of NGF used by Maffei and coworkers in their experiments, it is possible that the effects of NGF treatment in MD rats could be due to an "aspecific" action of NGF. Exogenously administered NGF could mimic the physiological action of a different neurotrophic factor endogenously produced by visual cortical neurons. It is likely that more than one neurotrophic factor could play a physiological role in the development and plasticity of the rat visual system.

In more recent experiments, Maffei and coworkers used transplanted Schwann cells as an exogenous source of neurotrophic factors to prevent the effect of monocular deprivation in the rat. Schwann cells are known to produce both NGF and other NGF-related molecules, such as Brain-derived Neurotrophic Factor (BDNF) (Bandtlow *et al.*, 1987; Acheson *et al.*, 1991). In these experiments, Schwann cells were transplanted in the lateral ventricles at postnatal day 14 and the animals were then monocularly deprived for one month. In the visual cortex, transplanted Schwann cells prevented the shift in ocular dominance and binocularity induced by monocular deprivation. In dLGN, shrinkage of neurons in the deprived laminae was also prevented (Pizzorusso *et al.*, 1994). Interestingly, the cotransplantation of hybridoma cells secreting an anti-NGF antibody (α-D11; Berardi *et al.*, 1994) counteracted, at least in part, the effects of transplanted Schwann cells in monocularly deprived rats (Pizzorusso *et al.*, 1994). The Authors concluded that transplanted Schwann cells prevented both physiological and anatomical effects of monocular deprivation presumably acting through the production of NGF. There remains open the possibility that other neurotrophins (directly or indirectly produced by Schwann cells) play a relevant role.

However, other recent results from the same group strongly suggest that NGF exerts a specific role in the maturation of geniculo-cortical connections in the rat. To test this hypothesis, Maffei and coworkers blocked the physiological action of NGF in the rat visual system by means of two distinct monoclonal antibodies (αD11 and 4C8) (Berardi et al., 1994). These antibodies were specific to NGF and did not block, at least in vitro, other members of the neurotrophin family. Therefore, αD11 or 4C8 antibody-secreting hybridoma cells were implanted into the lateral ventricle at postnatal day 15, and the effects of this treatment were evaluated at P45. Blocking in vivo the action of NGF resulted in dramatic alterations in the funtional properties of the visual system. A significant shrinkage in dLGN neurons was observed. Visual acuity was strongly reduced, and also binocularity of visual cortical neurons was markedly affected: cells in classes 1-3 were the great majority (~90%). These results are summarized in Fig. 1.3. Interestingly, rats implanted with αD11 hybridomas showed a prolonged period of sensitivity to monocular deprivation (critical period) (Domenici et al., 1994a). In rats transplanted from P14 to P45, monocular deprivation was still effective when performed for one month starting at P45. Therefore, the Authors postulated that the blockade of endogenous NGF could delay the process of synapse strengthening in the visual cortex.

A direct role of NGF in the development and plasticity of the rat visual cortex is also suggested by recent experiments on dark reared animals. As in higher mammals, dark rearing from birth markedly impairs the normal maturation and functional development of the visual cortex of the rat (Fagiolini et al., 1994a). In these experiments, rats were reared in complete darkness from birth and implanted at P13 with thermoplastic hollow capsules containing fibroblasts genetically engineered to produce NGF (Hoffman et al., 1993). Chronic treatment with NGF from P13 to P35-45 prevented the visual deficits induced by dark rearing. In these animals, many parameters of visual cortical function (orientation selectivity, low adaptation to repeated stimuli and visual evoked potentials -VEPs- in response to alternating gratings of various spatial and temporal frequency) were normal, as compared to dark reared animals (Pizzorusso et al., 1995). The same Authors previously demonstrated that also Schwann cells transplantation could prevent the effects of dark-rearing in the rat visual cortex (Fagiolini et al., 1994b).

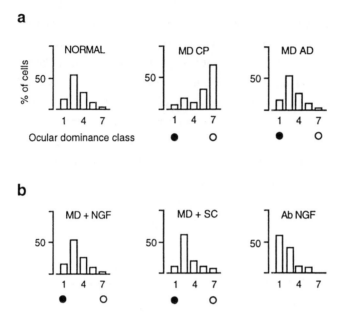

Figure 1.3.

Effect of monocular deprivation and NGF treatment on ocular dominance in the rat visual cortex.

a) In the adult rat visual cortex, the great majority of neurons is functionally connected with the contralateral eye (NORMAL). Monocular deprivation during the "critical period" (MD CP) causes a marked shift of this distribution in favor of the open eye (open circle), whereas monocular deprivation in adult animals (MD AD) has not any effect. Closed circle indicates the class of cells that respond to the deprived eye. Ocular dominance classes have been determined according to the classification by Hubel and Wiesel (Maffei *et al.*, 1992).

b) Exogenous administration of NGF or Schwann cell transplant in monocularly deprived rats (MD+NGF and MD+SC, respectively) prevents the shift in ocular dominance distribution caused by MD. Conversely, blocking *in vivo* the action of NGF by the transplant of anti-NGF antibody producing cells (Ab NGF) causes a marked loss of binocular neurons.

1.1.5   The role of other neurotrophic factors in the mammalian visual system.

The results obtained by Maffei and coworkers strongly supported the hypothesis that endogenous NGF is involved in the functional and anatomical development of the rat geniculo-cortical system. More recently, many results from other groups demonstrated that other neurotrophic factors of the NGF family could play a significant role in the maturation of connections in the cat visual cortex.

In their recent paper, Shatz and coworkers demonstrated that exogenous administration of Brain-derived Neurotrophic Factor (BDNF) or Neurotrophin-4/5 (NT-4/5), but not of NGF or Neurotrophin-3 (NT-3) inhibited the formation of ocular dominance columns in the cat (Cabelli *et al.*, 1995a). These Authors infused the visual cortex of cats with different neurotrophins during the critical period (from P28 to P42), and then studied the pattern of ocular dominance columns in the cortex by injecting tritiated proline in one eye. Animals treated with NGF, NT-3 or vehicle solution showed the classical alternated patches in the layer IV of primary visual cortex. Conversely, the treatment with BDNF or NT-4/5 completely disrupted this cortical columnar organization. The Authors postulated that both BDNF and NT-4/5 have a physiological role in the development of the visual cortex. This action would be mediated by their common receptor TrkB, that is present at the level of geniculo-cortical synapses (see below, in this Introduction). Preliminary unpublished results from the same group (see ref. 55 in Thoenen, 1995) confirmed this hypothesis: neutralizing antibodies against TrkB are able to block the formation of ocular dominance columns, when they are infused in the cat visual cortex during the critical period.

Very recent findings from the group of L. Katz confirmed a specific role of NT-4/5 in the maturation of geniculo-cortical connections in the ferret. As in other mammals, monocular deprivation in the ferret causes not only a reorganization of geniculo-cortical connections in the layer IV of primary visual cortex, but also determines a strong reduction in the soma size ("shrinkage") of geniculate neurons functionally connected to the deprived eye. Katz and coworkers studied the effect of intracortically applied neurotrophins in monocularly deprived ferrets. To this purpose, they developed a novel system to deliver neurotrophins *in vivo*, based on the intracortical injection of fluorescent latex microspheres coated with neurotrophins. The microspheres were taken up by the geniculate nerve terminals, and retrogradely transported to geniculate cell bodies. The use of fluorescent beads allowed an unambiguous detection of the geniculate neurons that were exposed to neurotrophins. By using this system, these Authors demonstrated that only NT-4/5 (and not NGF, NT-3 or BDNF) could rescue geniculate neurons from the shrinkage induced by MD (Riddle *et al.*, 1995).

In a series of *in vitro* experiments on slices from the ferret visual cortex, the same Authors also demonstrated that specific neurotrophins regulate the growth of dendritic arborization of defined populations of cortical neurons (McAllister *et al.*, 1995). Neurons in layer IV seem to respond most strongly to BDNF, whereas neurons in layers V-VI to NT-4/5. These results suggest that the TrkB ligands, BDNF and NT-4/5, could also play a critical role not only in the organization of intracortical connections in the visual cortex.

Taken together, results from the groups of Maffei, Shatz and Katz clearly demonstrate that neurotrophic factors of the NGF family are involved in the development and plasticity of the mammalian visual system. However, it is clear from these studies that different neurotrophins could exert different actions in the visual system of rats, cats and ferrets. Some possible explanations for these results will be discussed at the end of this review.

In the following sections of this Introduction, I will review the structure, function and mode of action of neurotrophic factors and their receptors, as well as their localization in the mammalian visual system. The mechanisms that regulate neurotrophin synthesis and release will be also discussed.

1.2   Neurotrophic factors of the NGF family.

Neurotrophic factors of the NGF family (also referred to as neurotrophins) are a class of structurally and functionally related molecules that regulate survival and differentiation of specific neuronal population of Peripheral and Central Nervous System. Their major functional properties, as resulting from classical studies and from the analysis of the recently generated neurotrophin-deficient mice, will be described here.

1.2.1   Nerve Growth Factor.

Nerve Growth Factor was originally discovered in the early 1950's for his selective growth-stimulating effects on the sensory and sympathetic nervous system (Levi-Montalcini, 1951). Few years later, Stanley Cohen purified NGF from mouse salivary gland and also provided the first experimental evidence that a neurotrophic factor is required during normal development: anti-NGF antibodies injected into newborn rodents destroyed the sympathetic chain ganglia (Choen, 1960). Subsequently, the amino acid sequence of NGF was determined (Angeletti and Bradshaw, 1971) and the cDNA was cloned (Scott et al., 1983; Ullrich et al., 1983). The mRNA for NGF encodes a precursor pre-pro-peptide of 252 aminoacids, that consists of three different subunits (α, β, and γ). Purified native NGF exists as complex of these three subunits, in a stoichiometry of $\alpha_2\beta_2\gamma_2$ (M_r, relative molecular weight: ~130,000). The immature pre-pro-peptide is cleaved by an endogenous proteolytical activity (located in the γ subunit) to give rise to the biologically active peptide, the β-homodimer (each β chain being 118 aminoacid long, M_r ~13,000). The three-dimensional structure of NGF protein has also been recently determined by X-ray crystallography (Mc Donald et al., 1991).

In the past fifteen years, many studies confirmed the original findings of Levi-Montalcini and Choen. At present, it is well known that NGF is produced in very low ("limiting") amounts by the targets of sensory and sympathetic neurons (see for example Bandtlow et al., 1987) and that it acts as a retrograde diffusible factor for the survival and differentiation of these neuronal populations in the Peripheral Nervous System (PNS) (reviewed in Levi-Montalcini, 1987, and Barde, 1989). A large series of studies also showed that NGF can act on specific neuronal populations in the CNS. In the brain, NGF is selectively produced by neocortex and hippocampus (Large et al., 1986), and it exerts a trophic action for striatal and septal cholinergic neurons of the basal forebrain that project to these structures. The first demonstration came from the observation that intracerebral injections of NGF can promote survival of septal cholinergic neurons after transection of the septo-hippocampal connection (that is commonly known to induce a marked degeneration of septal neurons) (Hefti, 1986; and for a review see also Thoenen et al., 1987).

1.2.2 *Brain-derived Neurotrophic Factor.*

The observation that only restricted populations of neurons both in the PNS and CNS are responsive to NGF, as well as the general assumption that many neurotrophic factors must account for the complexity of connectivity of the nervous system, led to the search and discovery of novel molecules related to NGF. In 1982, Brain-derived Neurotrophic Factor (BDNF) was purified from the pig brain (Barde et al., 1982) and its cDNA was subsequently cloned few years later

(Leibrock *et al.*, 1989). The deduced amino acid sequence revealed a high homology (50%) and a common structure with NGF. BDNF protein is synthesized as an immature polypeptide of ~250 residues (M_r: ~14,000), that is subsequently processed to give rise to the mature homodimeric peptide.

Despite to their similar structure, BDNF and NGF have very different distribution and range of action both in the PNS and CNS. In the PNS, BDNF mainly promotes the survival of neurons of the nodose ganglion (Hohn *et al.*, 1990) and trigeminal mesencephalic nucleus (Lindsay, 1988), but does not act on sympathetic ganglia (Lindsay *et al.*, 1985; Leibrock *et al.*, 1989; Maisonpierre *et al.*, 1990a). Specific subpopulations of dorsal root ganglia (DRG) neurons are also supported by the action of BDNF (Lindsay, 1988; Leibrock *et al.*, 1989). In the CNS, BDNF is highly expressed in the neocortex, hippocampus, and cerebellum (Hofer *et al.*, 1990; Ernfors *et al.*, 1990; Phillips *et al.*, 1990). It has been calculated that in these brain regions, the level of BDNF mRNA is at least 50-fold higher than that of NGF mRNA (Hofer *et al.*, 1990). Midbrain (optic tectum), striatum and spinal cord also produce, at lower levels, the BDNF mRNA (Hofer *et al.*, 1990). A recent himmunohistochemical analysis also revealed a comparable localization of the BDNF protein in the brain (Dugich-Djordjevic *et al.*, 1995). This wide distribution of BDNF in the CNS clearly resembles its spectrum of action. A large series of studies has demonstrated that mesencephalic dopaminergic neurons of the substantia nigra (Hyman *et al.*, 1991) and magnocellular cholinergic neurons of the basal forebrain (Alderson *et al.*, 1990; Knüsel *et al.*, 1992), as well as retinal ganglion cells (Rodriguez-Tébar *et al.*, 1989) and motoneurons (DiStefano *et al.*, 1992) are all responsive to BDNF.

1.2.3   Other factors related to NGF and BDNF.

After the molecular cloning of BDNF, several other related factors have been cloned by using polymerase chain reaction (PCR) with primers corresponding to the conserved sequences of NGF and BDNF. Neurotrophin-3 (NT-3: Hohn *et al.*, 1990; Maisonpierre *et al.*, 1990a), Neurotrophin-4/5 (NT-4/5: alternatively known as NT-4 or NT-5) (Berkemeier *et al.*, 1991; Hallböök *et al.*, 1991; Ip *et al.*, 1992) and the recently isolated Neurotrophin-6 (NT-6: Götz *et al.*, 1994) all share a ~50% homology with NGF and BDNF, and have some other important common features. They are all synthesized as immature precursors that undergo proteolytic cleavage to give rise to the biologically active molecules (M_r: ~13,000). The amino-terminal of the mature protein contains a signal peptide sequence, as expected for secretory proteins. Figure 1.4 A shows a schematic representation of the neurotrophin structure.

Among these members of the NGF/BDNF family, NT-3 and NT-4/5 are the best characterized. The spectrum of action of these factors is restricted to specific subsets of neuronal populations and is, at least in part, overlapping to that of NGF and BDNF. For example, NT-3 is a specific survival factor for motoneurons (DiStefano *et al.*, 1992), and also acts on neurons of nodose (Rosenthal *et al.*, 1990) and dorsal root (Maisonpierre *et al.*, 1990a) ganglia. In the brain, very low levels of NT-3 mRNA have been reported in the hippocampus (Phillips *et al.*, 1990), substantia nigra (Seroogy *et al.*, 1994) and cerebellum (Maisonpierre *et al.*, 1990b), but its trophic action on specific neurons of the CNS has not been clearly demonstrated so far.

NT-4/5 supports the survival of DRG and sympathetic neurons in the PNS (Berkemeier *et al.*, 1991), whereas no clear evidence has been reported for a role of this neurotrophin in the CNS. However, the widespread distribution of NT-4/5 in different brain regions (hippocampus, neocortex, thalamus and cerebellum; Timmusk *et al.*, 1993a) seems to suggest a trophic action for this factor on multiple neuronal populations in the CNS.

1.2.4   Neurotrophin-deficient mice and "neurotrophic hypothesis".

All these studies strongly confirmed the idea that neurotrophins act as survival factors for specific subsets of neuronal populations. During development, neurons are generated in large excess, and are eliminated by an active process of naturally occurring (programmed) cell death that takes place during late embryonic and early postnatal life. The number of neurons in the adult nervous system depends on a competitive process in which the innervated tissues produce limiting amounts of trophic factors that support neuronal survival. Only those neurons that receive enough neurotrophic factor can establish the most effective synapses with their target tissues, and therefore survive. This theory, generally known as "neurotrophic hypothesis" (reviewed in Purves, 1986, and Oppenheim, 1991), has been recently confirmed by several Authors who used homologous recombination to generate mutant mice lacking specific neurotrophins (reviewed in Klein, 1994 and Snider, 1994). For example, sympathetic chain ganglia are virtually absent in newborn mice lacking NGF (Crowley *et al.*, 1994). Moreover, these animals show a strong reduction in the number of nociceptive and thermal receptive DRG neurons and therefore fail to respond to noxious mechanical stimuli. The most striking abnormality in mice lacking BDNF (Ernfors *et al.*, 1994a; Jones *et al.*, 1994) is the almost complete (>80%) loss of neurons of the vestibular ganglion, that determines an abnormal locomotor behavior. Neurons of trigeminal and dorsal root ganglia are also significantly depleted in BDNF (-/-) mutant mice. In mice lacking NT-3 (Ernfors *et al.*, 1994b; Fariñas *et al.*, 1994) the complete loss of proprioceptive neurons of DRG and muscle spindles is the primary cause of their impaired locomotor activity (limb ataxia and athetotic walking movements). Finally, mice lacking NT-4/5 (Conover *et al.*, 1995; Liu *et al.*, 1995) show no obvious neurological defects but exhibit a loss of sensory neurons of nodose-petrosal ganglion.

Taken together, all these data strongly support the idea that neurotrophic factors act as survival factors for specific neuronal poulations of the PNS. Moreover, mice heterozygous for neurotrophin gene deletions show reduced physiological and anatomical abnormalities as compared to their homozygous littermates, thus confirming the hypothesis that neurotrophic factors act in limiting amounts *in vivo*. However, these studies did not shed, at least so far, sufficient light onto the "survival-promoting" action of neurotrophins in the CNS. In the PNS, most neurons innervate a limited number of target cell types, and receive no innervation (sensory ganglion cells) or innervation from only one source (sympathetic ganglion cells). In contrast, many CNS neurons establish a huge number of synaptic contacts, projecting to multiple targets and receiving multiple innervation. The study of trophic interactions between CNS neurons results therefore to be extremely difficult, as compared to the PNS. For example, the finding that survival of magnocellular cholinergic and nigral dopaminergic neurons is not affected in mice lacking NGF and BDNF, respectively, suggests that compensatory events could take place in the absence of one of these factors and confirms the hypothesis of a multiple action of neurotrophic factors on specific neuronal populations in the developing CNS. Moreover, there is still the possibility that neurotrophic factors regulate specific CNS functions different than survival. Neurotrophin-deficient homozygous mice generally die soon after birth (unless NT-4/5 -/- mutant mice), thus not allowing any extensive anatomical and physiological investigation in the adult CNS. The analysis of heterozygous mice, that are viable and show no obvious behavioral abnormalities, could be therefore useful to address this point. In particular, those functions that are supposed to be dependent, also in the adult CNS, on limiting amount of neurotrophins (such as synaptic strengthening and maintenance) could be adequately studied in these animals. In fact, recent results suggest that neurotrophic factors could actively involved in neuronal plasticity in the CNS. These data have been recently reviewed by Thoenen (1995) and will be discussed in Chapter 3.

1.3   Neurotrophin receptors.

Neurotrophic factors are secretory proteins that exert their action on target cells through the binding to specific transmembrane receptors. The structure and function of these receptors have been extensively reviewed by several Authors (Meakin and Shooter, 1992; Glass and Yancopoulos, 1993; Raffioni et al., 1993; Barbacid, 1995; Bothwell, 1995; Chao and Hempstead, 1995), to whom the reader is referred for additional details.

1.3.1   Two classes of NGF receptors.

Early studies (Fraizer et al., 1974; Sutter et al., 1979; Schechter and Bothwell, 1981; Godfrey and Shooter, 1986) revealed that a variety of NGF-responsive neurons and cultured cells displayed two different classes of NGF receptors: a low-affinity ($K_d = 10^{-8}$ - 10^{-9} M) and a high-affinity ($K_d = 10^{-10}$ - 10^{-11} M) receptor. These two receptor classes could be distinguished by some important features. The low-affinity NGF receptor (LNGFR) was trypsin-labile and had a fast dissociation rate for NGF ($t_{1/2}$ of approximately 3 s), while the high-affinity NGF receptor (HNGFR) was trypsin-stabile and slow-dissociating ($t_{1/2}$ of approximately 10 min). Chemical cross-linking experiments (Grob et al., 1983) also demonstrated that LNGFR and HNGFR had an apparent molecular wheight of approximately 75 kDa and 135-140 kDa, respectively. Interestingly, some of these early studies demonstrated that only HNGFR was biologically active in NGF-responsive cells (Hosang and Shooter, 1985) and could internalize bound NGF (Bernd and Greene, 1984).

1.3.2   p75, the low-affinity neurotrophin receptor.

The purification of LNGFR (Puma et al., 1983), and the subsequent availability of monoclonal antibodies against the receptor molecule (Chandler *et al.*, 1984), allowed the molecular cloning of both human and rat LNGFR cDNAs (Chao *et al.*, 1986; Radeke *et al.*, 1987). The primary structure of LNGFR was determined from the corresponding nucleotidic sequence. The mature protein contains 397-399 amino acids (plus a signal peptide sequence), with an apparent molecular weight of approximately 75 kDa. The extracellular domain (~220 amino acids at the amino-terminal) contains a cystein-rich motif that is supposed to be involved in NGF binding (Baldwin *et al.*, 1992). The cytoplasmic domain (~150 amino acids) lacks any consensus sequence for kinase or any other catalytic activity.

Binding experiments on LNGFR-transfected cell lines demonstrated that this receptor can bind all neurotrophins with similar (low) affinity (Rodriguez-Tébar *et al.*, 1990, 1992; Hallböök *et al.*, 1991; Squinto *et al.*, 1991). Therefore, this receptor is commonly referred to as p75-low affinity neurotrophin receptor, or, in a simpler way, p75. However, several studies demonstrated that p75-transfected fibroblasts do not respond to NGF, nor they exhibit high affinity binding sites. These observations suggested that another receptor protein might be involved in neurotrophin signalling.

1.3.3   *The Trk family of high-affinity neurotrophin receptors.*

The *Trk* oncogene was originally discovered as a tropomyosin gene-associated sequence in human colon carcinomas (Martin-Zanca *et al.*, 1986). Sequence analysis of the *Trk* proto-oncogene (successively named TrkA) revealed that the gene encodes for a putative tyrosine-kinase receptor protein (Martin-Zanca *et al.*, 1989). The apparent molecular weight of the mature protein (790 amino acids) is 140 kDa. The extracellular domain contains multiple putative N-glycosylation sites and two immunoglobulin IgG-like motifs, while the tyrosine-kinase catalytic activity is located in the intracellular domain.

The original observation that TrkA mRNA was expressed in NGF-responsive neuronal sensory ganglia (Martin-Zanca *et al.*, 1990) suggested that TrkA could act as a receptor for NGF. Further studies demonstrated that TrkA is the high-affinity NGF receptor (Kaplan *et al.*, 1991a, b; Klein *et al.*, 1991a).

Low-stringency screenings of brain cDNA libraries with TrkA probes led to the isolation of cDNA clones with structural homologies to TrkA. Both TrkB (Klein *et al.*, 1989) and TrkC (Lamballe *et al.*, 1991) code for tyrosine-kinase receptors that have ~50% and ~80% amino acid identity with TrkA in their extracellular and intracellular domains, respectively. The apparent molecular weight of these two proteins is appoximately 140 kDa. The TrkB cDNA codes for a functional high-affinity receptor for BDNF and NT-4/5 (Berkemeier et al., 1991; Klein et al., 1991b; Ip et al., 1992), while the product of the TrkC cDNA specifically binds NT-3 (Lamballe et al., 1991). NT-3 can also bind, with lower affinity, both TrkA and TrkB. Figure 1.4 B and C respectively show the structure of neurotrophin receptors and their binding specificity to the different neurotrophins.

A large series of in situ hybridisation and immunohistochemical studies contributed in the past recent years to characterize the distribution of neurotrophin receptors in the PNS and CNS. In the PNS, neurotrophin receptors are highly expressed in specific neuronal populations of the ganglia that innervate neurotrophin-producing peripheral organs (reviewed in Klein, 1994; Snider, 1994; Barbacid, 1995). In the CNS, TrkA and p75 are mainly localized in cholinergic neurons of the basal forebrain that project to hippocampus and neocortex (Gibbs and Pfaff, 1994; Sobreviela et al., 1994), while TrkB and TrkC are widely distributed throughout the brain (neocortex, hippocampus, thalamus, ventral midbrain and cerebellum: Merlio et al., 1992; Ringstedt et al., 1993).

1.3.4   Truncated isoforms of Trk receptors.

Alternatively spliced isoforms for all the members of the Trk family have been described by several Authors (Klein et al., 1990; Middlemas et al., 1991; Valenzuela et al., 1993; Clary and Reichardt, 1994). So far, no clear demonstration of their function has been given. However, some different hypotheses have been proposed. (1) Truncated Trk isoforms lacking the catalytic kinase domain could reduce responsivess to neurotrophin by either acting as dominant negative effector of full-length Trk isoform or as nonfunctional receptors for neurotrophins. (2) Truncated Trks could have adhesive properties due to their limited homology to known cell adhesion molecules. (3) Finally, they could play a role in the presentation of neurotrophins to full-length receptors.

1.3.5   Proposed models for the biologically active neurotrophin receptor.

After the discovery of the Trk tyrosine kinases as high-affinity receptors for neurotrophins, a large series of studies aimed to elucidate the role of Trks and

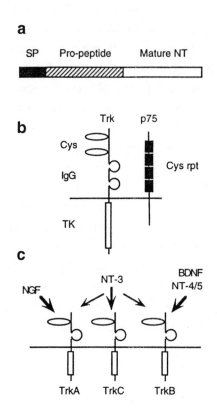

Figure 1.4.

Neurotrophins and their receptors.

a) Structure of neurotrophic factors. Neurotrophins (NT) are synthesized as immature pre-pro-peptides and then processed by proteolytic cleavage to give the mature (biologically active) peptide. SP (black box), signal peptide sequence. b) Trk and p75 neurotrophin receptors. TrkA, B and C all have a common structure. Cys, cysteine rich domains. IgG, immunoglobulin-like domains. TK, tyrosine-kinase domain. Truncated isoforms (not shown in this figure) lack the TK domain. p75 has not any catalytic intracellular domain. Cys rpt, cysteine repeats. c) Binding interactions between neurotrophins and Trks. (Modified from Thoenen, 1995).

p75 in the specification of biologically active neurotrophin receptors. The review of all these studies is certainly beyond the purpose of this Thesis. The reader is therefore referred to Bothwell (1995) and Chao and Hempstead (1995) for an extensive discussion on the recent advances in this field.

It is generally accepted that Trks can act alone as high-affinity neurotrophin receptors. Neurotrophin binding induces Trk homodimerization and trans-autophosphorylation, thus initiating the signalling cascade (see for example Jing et al., 1992, and section 1.6 for further details), and this step does not require p75. Moreover, Trk-deficient mice, unlike p75 knockouts, exhibit patterns of neuronal cell loss that overlap to those observed in the corresponding neurotrophin-deficient animals (see below). However, monoclonal antibodies against p75 can decrease the NGF-induced autophosphorylation of TrkA (Barker and Shooter, 1994), thus suggesting that binding of NGF to p75 can affect TrkA signalling.

Two hypothetical models for p75-TrkA interactions have been proposed (Chao and Hempstead, 1995), that could also be applied to the other neurotrophin receptors. The "presentation model" postulates that the fast binding of NGF to p75 could increase the local concentration of NGF for TrkA. According to the "conformational model", the co-expression of both receptors would lead to a high-affinity binding site by inducing a conformational change in TrkA. This latter model would be consistent with much of the available data.

Another interesting model for the action of p75 has been proposed by Bothwell (1995): p75 could act in facilitating retrograde signalling of neurotrophin-responsive projection neurons. This hypothesis rises from the observation that both in the PNS and CNS p75 is mainly expressed in projection neurons. For example, in the basal forebrain, both cholinergic neurons of the septo-hippocampal projection and cholinergic interneurons of the striatum express TrkA and respond to NGF, but only the former express p75 (Steininger et al., 1993; Sobreviela et al., 1994). Projection neurons, unlike interneurons, would require retrograde neurotrophin transport to obtain their trophic support, and p75 could be therefore involved in this action. The recent finding that the retrograde transport of BDNF and NT-4/5 in peripheral neurons is reduced in p75-deficient animals (Curtis et al., 1995) is consistent with this model.

More recently, it has been shown that p75 alone is sufficient to transduce neurotrophin signalling in specific cell types. In a glioma cell line, neurotrophin binding to p75 activates the sphingomyelinase pathway for the production of the lipid messenger ceramide (Dobrowsky *et al.*, 1994). In rat Schwann cells, that do not express TrkA, NGF (but not BDNF or NT-3) binding to p75 leads to the activation of the transcription factor NF-κB (Carter *et al.*, 1996). Thus, p75 seems to be involved, at least in some specific cell types, in signal transduction mechanisms that are indipendent of Trk proteins.

1.3.6 *p75 and Trks knockouts.*

In the past three years, several groups have generated mice lacking neurotrophin receptors by using homologous recombination (reviewed in Klein, 1994, and Snider, 1994). The phenotypes of Trk-deficient animals are in a good agreement with those obtained with mice lacking the corresponding neurotrophins. As already observed in neurotrophin-deficient animals, Trk knockouts exhibit marked alterations mainly in the PNS but not in the CNS. Moreover, mice heterozyigous for neurotrophin receptor mutations have, as expected, a weaker phenotype as compared to their homozygous littermates. In TrkA-deficient animals (Smeyne *et al.*, 1994) the complete loss of sensory (nociceptive) neurons is the primary cause of their reduced response of painful stimuli, as already observed in NGF-deficient mice. Sympathetic ganglia are also completely destroyed in these animals, while no apparent alterations can be observed in the cholinergic nuclei of the basal forebrain. Mice lacking the receptor for BDNF and NT-4/5 (TrkB) (Klein *et al.*, 1993) exhibit a more severe phenotype, as compared to those of BDNF, NT-4/5 or BDNF/NT-4/5-deficient animals (Conover *et al.*, 1995; Liu *et al.*, 1995). These results suggest that another ligand, perhaps NT-3, could act on TrkB *in vivo*. TrkB-deficient mice generally die during the first postnatal days, and have an almost complete (80%) as well as a significant (35%) loss of vestibular ganglia and motor neurons, respectively. Finally, TrkC mutant mice (Klein *et al.*, 1994) show the same strong reduction of proprioceptive DRG neurons as observed in NT-3 knockouts, thus confirming the hypothesis that proprioceptors are dependent on NT-3.

Surprisingly, the phenotype of p75-deficient animals (Lee *et al.*, 1992, 1994) is only partially overlapping to that previously described for NGF-deficient mice, and has no similarities to the phenotype of other neurotrophin- or Trk-deficient mice. In fact, these animals have a deficit of nociceptive function and lack sympathetic innervation of pineal and sweat glands. This results suggest a functional role for p75 in the development of restricted NGF-responsive neuronal populations.

Results obtained with neurotrophin receptor-deficient mice strongly confirmed that Trk receptors mediate the action of neurotrophins on specific neurons of the PNS, but failed to give clear evidence (except for TrkA -/- mice) on their role in the CNS. The analysis of TrkA -/- mice revealed a marked reduction of acetylcholinesterase in the target areas of the TrkA-positive cholinergic neurons of the basal forebrain (hippocampus and neocortex; Smeyne *et al.*, 1994), thus confirming a possible physiological role of TrkA in these brain regions. Conversely, TrkB- and TrkC-deficient homozygous animals show no apparent alteration of brain structure. However, these animals generally die soon after birth, thus not allowing any extensive anatomical and physiological investigation in the CNS. For these reason, the analysis of mice heterozygous for TrkB and TrkC mutations will be probably useful for a better understanding of the function of these two neurotrophin receptors. Moreover, another possibility remains open. Some neuronal populations in the CNS are known to express more than one type of Trk receptors: for example, cholinergic neurons of the basal forebrain contain both TrkA and TrkB, whereas nigral dopaminergic neurons express both TrkB and TrkC (reviewed in Lindsay *et al.*, 1994). Therefore, in Trk-deficient animals, compensatory events could take place in those neurons that express more than one Trk receptor (i.e., up-regulation of TrkB in the absence of TrkC or vice-versa). Further investigation is needed to address this point.

1.4   Distribution of neurotrophins and their receptors in the mammalian visual system.

Since the appearence of the first papers of the group of L. Maffei, that proposed a "neurotrophic hypothesis" for the development of geniculo-cortical connections in mammals (Domenici *et al.*, 1991; Maffei *et al.*, 1992; see above, in this Introduction), several Authors have investigated the distribution of neurotrophic factors of the NGF family in the mammalian visual system. Here I will focus on those results that refer to the expression of neurotrophins and their receptors in the geniculo-cortical pathway of the visual system. These results are summarized in Table 1.

1.4.1   NGF,TrkA and p75.

NGF protein and mRNA are present in the rat neocortex with a peak of expression at postnatal day 21 (Large *et al.*, 1986). In particular, NGF protein and mRNA have been detected also in the visual cortex, where their peak of expression seems to correspond to peak of the critical period for the development of geniculo-cortical connections (Hayashi *et al.*, 1990; Schoups *et al.*, 1995; Bozzi, Cremisi, Pizzorusso and Maffei, unpublished observations). NGF mRNA was also detected in dLGN (Schoups *et al.*, 1995), where it could act as a survival factor for retinal ganglion cells that send their projections to the dLGN and are known to express both p75 and TrkA (Carmignoto *et al.*, 1991; Zanellato *et al.*, 1993). On the contrary, dLGN neurons do not express TrkA mRNA (Holtzman *et al.*, 1992; Merlio *et al.*, 1992; Schoups *et al.*, 1995) and do not retrogradely transport iodinated NGF when it is injected into the visual cortex (as the cholinergic afferents from basal forebrain do; Domenici *et al.*, 1994b).

The presence of TrkA in the neocortex is still controversial. Some Authors reported that TrkA mRNA is expressed in the rat cortex (Miranda *et al.*, 1993; Valenzuela *et al.*, 1993), and results obtained by reverse-transcriptase PCR (RT-PCR) on RNA extracted from the visual cortex confirm this finding (Cellerino and Maffei, 1996). Other Authors failed to detect any signal by *in situ* hybridisation (Gibbs and Pfaff, 1994) and Schoups *et al.* (1995) could not detect TrkA mRNA in the rat visual cortex also by using a sensitive RNAase protection assay. So far, it has not been clearly demonstrated that TrkA protein is present in the visual cortex. For example, immunohistochemical analysis with a TrkA-specific antibody (Clary *et al.*, 1994) failed to detect TrkA immunoreactivity in the rat cortex (Sobreviela et al., 1994). Conversely, immunoblot analysis with the same anti-TrkA antibody has recently demonstrated that TrkA is present in the rat visual cortex (Rossi, Viegi, Pizzorusso and Maffei, unpublished results). Therefore, it remains to be determined if cortical neurons do really contain TrkA. In fact, it can not be excluded that TrkA protein is localized on axon terminals of the cholinergic neurons of the basal forebrain, that are known to produce both TrkA mRNA and protein (Gibbs and Pfaff, 1994; Sobreviela *et al.*, 1994), and innervate all the neocortex, including the visual cortex (Carey and Rieck, 1987).

The low affinity neurotrophin receptor p75 is not expressed in the geniculo-cortical pathway. In a large series of *in situ* hybridisation experiments, we failed to detect p75 mRNA in the rat geniculate nucleus and neocortex (Bozzi,

Cremisi, Pizzorusso and Maffei, unpublished results). p75-immunoreactive fibers have been detected in the rat cortex (Pioro and Cuello, 1990), but they correspond to axon terminals of the cholinergic neurons of the basal forebrain, that project to the neocortex.

1.4.2   BDNF and TrkB.

So far, different studies demonstrated that BDNF mRNA and protein are also present in the rat visual cortex and increase with the beginning of the critical period for the development of the geniculo-cortical connections (Castrén et al., 1992; Bozzi et al., 1995; Schoups et al., 1995; Rossi et al., 1996).

The mRNA for BDNF/NT-4/5 receptor (TrkB) is also transiently expressed in dLGN and visual cortex during the first postnatal weeks (Castrén et al., 1992; Ringstedt et al., 1993; Bozzi et al., 1995; Schoups et al., 1995). Moreover, Allendoerfer et al. (1994) demonstrated that in the ferret visual system, the amount of truncated and full-length TrkB is developmentally regulated. By using cross-linking experiments to iodinated BDNF, these Authors showed that during the period of cortical segregation into eye-specific layers (first postnatal weeks), the amount of truncated TrkB in dLGN and cortex markedly increases relative to full-length isoform. The same results have also been recently obtained for mRNAs encoding the truncated and full-length isoforms of TrkB (Schoups et al., 1995). Taken together, these results strongly suggest that geniculo-cortical synapses, during the critical period of segregation, express a functional TrkB receptor.

1.4.3   NT-3, NT-4/5 and TrkC.

In the rat visual cortex, NT-3 mRNA is highly expressed at birth and then declines to undetectable levels before the time of natural eye-opening (Schoups et al., 1995; Lein et al., 1995). In situ hybridisation revealed that in the cat and ferret visual cortex, NT-3 mRNA is mainly expressed in layer IV (Lein et al., 1995). Layer IV of the visual cortex also specifically express, during the critical period, NT-4/5 mRNA (in the mouse; Bozzi, unpublished results) as well as NT-4/5 protein (in the cat and ferret; Cabelli et al., 1995b). This pattern of NT-4/5 expression is consistent with the observation that this neurotrophin can interfere with the formation of ocular dominance columns (Cabelli et al., 1995a, b). NT-4/5 mRNA has been recently detected also in the mouse dLGN by in situ hybridisation

(Bozzi, unpublished results).

Finally, both TrkC mRNA and protein are present in the rat and ferret visual cortex and dLGN (Allendoerfer *et al.* 1994; Schoups *et al.*, 1995).

		dLGN	VISUAL CORTEX
NGF	mRNA	+	+
	protein	ND	+
BDNF	mRNA	-	+++ (layers 2-3 and 5-6)
	protein	-	+++
NT-3	mRNA	+	+ (layer 4)
	protein	ND	ND
NT-4/5	mRNA	+++	++ (layer 4)
	protein	ND	++ (layer 4)
TrkA	mRNA	-	?
	protein	-	+ (cholinergic terminals?)
TrkB	mRNA	++	++ (layers 2-6)
	protein	++	++
TrkC	mRNA	+	+
	protein	+	+
p75	mRNA	-	-
	protein	-	+ (cholinergic terminals)

Table 1. Expression of neurotrophins and their receptors in the mammalian geniculo-cortical system. Abbreviations used in the table: -, not present. +/++/+++, low, medium and high level of expression (arbitrary units). ND, not determined. ?, contrasting results have been obtained from different groups. See text for references.

1.5   Mechanism of action of neurotrophic factors.

The intracellular signalling cascade that leads to both short- and long-term changes in response to neurotrophin stimuli has been widely investigated by several Authors (for extensive reviews, see Schlessinger and Ullrich, 1992; Raffioni *et al.*, 1993; Heumann, 1994; Greene and Kaplan, 1995; Segal and Greenberg, 1996). NGF-responsive PC12 cells** (that are known to express TrkA: Kaplan *et al.*, 1991b) and, more recently, Trk-transfected cell lines, have been used as experimental models for understanding the mode action of neurotrophic factors. Therefore, most of our knowledge in this field concerns the signal transduction cascade activated by NGF, while the characterization of other neurotrophin pathways is less advanced. Few data are also available for neurotrophin signalling in primary cultures or *in vivo*. The most important results of these studies will be discussed in the following sections.

1.5.1   Activation of Trk receptors.

Neurotrophic factors exert their biological action on target cells through the binding to Trk receptors. Following the binding of neurotrophin dimers, Trk receptors immediately undergo homodimerization. Ligand-induced receptor homodimerization is a general mechanism used by receptor tyrosine kinases (RTKs) for initiating the signal transduction cascade in response to growth factors (Schlessinger and Ullrich, 1992). Receptor homodimerization induces a rapid conformational change that leads to the activation of the intrinsic tyrosine kinase activity and to trans-autophosphorylation of the intracellular catalytic domain. Therefore, neurotrophin binding is immediately followed by Trk tyrosine-phosphorylation. This mechanism has been clearly demonstrated for the activation of TrkA (Kaplan *et al.*, 1991b; Jing *et al.*, 1992) and TrkB (Klein *et al.*, 1991b; Middlemas *et al.*, 1994) in response to NGF and BDNF, respectively.

**PC12 cells (Greene and Tischler, 1976) are a clonal cell line derived from a tumor of chromaffin cells of the rat adrenal medulla (pheochromocytoma). These cells, which do not require NGF for their survival and growth, acquire and maintain a sympathetic-like neuronal phenotype in response to NGF. This neuronal-like phenotype is characterized by neurite outgrowth and *de novo* expression of neurospecific markers and functions (such as neurofilaments, neurotransmitter synthesis and storage, neuronal excitability).

The time-course of neurotrophin-induced Trk tyrosine phosphorylation has been extensively studied. *In vitro*, TrkA phosphorylation is rapidly induced (within 1-5 minutes after NGF treatment) both on PC12 cells (Kaplan *et al.*, 1991b) or in freshly dissected tissues (septum, striatum and neocortex; Knüsel *et al.*, 1994), and lasts for at least 1-3 hours. *In vivo*, TrkA phosphorylation of cholinergic septal neurons induced by a single intracerebral injection of NGF seems to be longer lasting (up to 48 hours; Li *et al.*, 1995).

Tyrosine autophosphorylation sites have been mapped on both TrkA and TrkB receptors. The major phosphorylation sytes are located in the region containing tyrosine residues Y670, Y674 and Y675, in the kinase domain (Middlemas *et al.*, 1994; Stephens *et al.*, 1994). These residues are the first to be phosphorylated after neurotrophin binding. Subsequently, the juxtamembrane residue Y490 and carboxy-terminal region residue Y785 undergo phosphorylation: this step is crucial for the activation of downstream signalling cascade. (Stephens *et al.*, 1994).

1.5.2   *Signal transduction cascades.*

Autophosphorylation events described above are a general mechanism used by growth factor tyrosine kinase receptors (such as EGF, PDGF or neurotrophin receptors) to activate signal transduction. Phosphorylation of tyrosine residues creates new binding sites for proteins carrying the src-homology motif (SH2), a domain of about 100 amino acids with high affinity to phosphotyrosine (P-Tyr) residues located within defined sequences (see Schlessinger, 1994, for a review). Following their binding to P-Tyr residues of the receptor, SH2 domain proteins are themselves phosphorylated and activate specific kinase pathways.

Tyrosine residues Y490 and Y785 on both TrkA and TrkB are specific binding sites for the SH2 domain-containing proteins SHC and phospholipase C-γ1 (PLC-γ1), respectively (Middlemas *et al.*, 1994; Obermeier *et al.*, 1994; Stephens *et al.*, 1994). After its binding to TrkA, SHC becomes phosphorylated and associates with the cytoplasmic complex GRB2-SOS (Stephens *et al.*, 1994). The GRB2-SOS complex is therefore recruited to the plasma membrane, and this event leads to the activation of the small G protein p21ras: SOS ("Son of Sevenless") protein is an exchange factor that promotes the binding of GTP to p21ras (reviewed in McCormick, 1994). The activation of p21ras starts a downstream phosphorylation cascade that allows the neurotrophin signal to be trasferred to the cell nucleus (reviewed in Davis, 1993; Heumann, 1994; Segal and Greenberg, 1996). The chain of phosphorylation events initiates with the activation of the serine/threonine kinase *Raf* by p21ras. Furthermore, the mitogen-activated protein kinases MAPKK and MAPK are activated. MAPK consists of two related kinases, the extracellular signal-regulated kinases ERK1 and ERK2. Following MAPK phosphorylation, different transcription factors become activated: ERK2 and ERK1 have been shown to phosphorylate the transcription factors myc and Elk, respectively. Elk binds with Serum Response Factor (SRF) to the c-fos promoter, thus initiating the immediate early gene response.

The ras-raf-MAPK cascade described above is the best understood signalling pathway activated by NGF in PC12 cells (reviewed in Heumann, 1994 and Segal and Greenberg, 1996). However, other alternative (ras-indipendent) pathways have been proposed. For example, very rapid modulation of phospholipid synthesis has been reported in PC12 cells treated with NGF (reviewed in Segal and Greenberg, 1996). This could be ascribed to the activation of phospholipase C-γ1 (PLC-γ1) that occurs following TrkA tyrosine-phosphorylation (see above). Further evidence for NGF-induced activation of the phospholipid-diacylglicerol-protein kinase C pathway is still lacking. Interestingly, NGF has been shown to induce a direct association between ERK1 and TrkA (Loeb *et al.*, 1992), thus amplifying the number of possible membrane targets that are phosphorylated by ERK1 in response to NGF. Finally, an NGF-specific and ras-dependent kinase has been characterized in PC12 cells that stimulates c-fos transcription via phosphorylation of CREB (cAMP responsive element binding protein, see 1.5.3). This new kinase has been named CREB kinase (Ginty *et al.*, 1994). All these different pathways are summarized in Figure 1.5, that illustrates a possible model for the Trk-mediated neurotrophin signal transduction cascade. Finally, some different Trk-independent signaling mechanism that involve the p75 low affinity neurotrophin receptor have been also described (Dobrowsky *et al.*, 1994; Carter *et al.*, 1996). This findings have been already reported above in this Introduction.

1.5.3 Neurotrophin-induced gene expression.

The phosphorylation cascade activated by NGF leads to the activation of specific subsets of transcription factors that regulate the expression of a large number of genes in PC12 cells. These genes have been classified in two different classes: immediate early genes (IEGs) and late response genes (LRGs) (see Sheng and Greenberg, 1990 and Segal and Greenberg, 1996 for a review). Induction of IEGs by NGF is rapid, transient (it appears within minutes and lasts less that one hour) and independent of new protein sysnthesis. Many NGF-inducible IEGs code for transcription factors, whose expression is required for the delayed induction of LRGs. Late response genes are slowly regulated, over a time frame of hours, and their products act as specific effectors for the acquisition of the NGF-induced neuronal phenotype of PC12 cells.

Among NGF-induced IEGs, the proto-oncogene c-fos is the best characterized. Fos belongs to the Fos-Jun family of "leucine-zipper" transcription factors (reviewed in Sheng and Greenberg, 1990 and Morgan and Curran, 1991). In PC12 cells, NGF induces c-fos transcription via a specific mechanism that requires the cooperative interaction of both SRF and CREB transcription factors (see above) on the c-fos promoter (Bonni et al., 1995). Thus, the two distinct kinase pathways that indipendently lead to the activation of SRF and CREB seem to converge on the c-fos promoter to confer specificity to NGF signal (in other systems, phosphorylated CREB can mediate calcium- or cAMP-induced c-fos transcription independently of other promoter-bound transcription factors; see Sheng et al., 1990, and Ginty et al., 1993). Many other NGF-induced transcription factors have been characterized: c-myc, zif/268 (NGF-IA), nur 77 (NGF-IB) all respond to NGF in PC12 cells (reviewed in Sheng and Greenberg, 1990). Signal transduction cascades that lead to their activation are not well known as the c-fos pathway. However, the possibility that multiple pathways for NGF signalling may exist in PC12 cells is currently under investigation.

In PC12 cells, late response genes encode a large number of neural-specific markers. Among them, I can mention different subunits of neurofilaments (NF-L, NF-M), neural cell adhesion molecules (N-CAM), vesicle-associated proteins (such as SCG10), neuropeptide-like molecules (VGF) and sodium channels (brain type II/IIA and perypheral nerve type 1, PN1) (for references see Sheng and Greenberg, 1990; Levi and Alemà, 1991). Their expression, slowly induced and sustained by NGF, contributes to the acquisition of specifc neuronal properties such as neurite outgrowth and membrane excitability. The onset of some specific NGF-induced differentiation programs seems to be very rapid and selective. For example, it has recently been demonstrated that a single, brief exposure (1 min) of PC12 cells to NGF is sufficient to "trigger" long-term

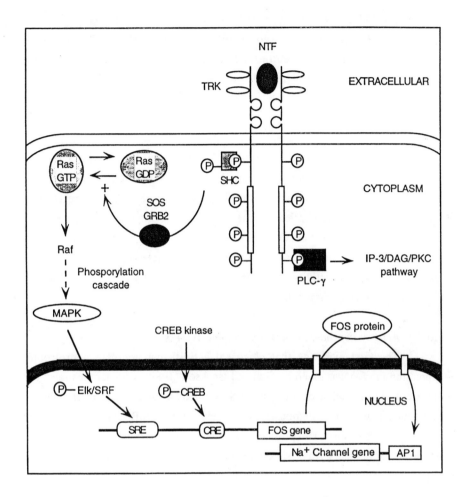

Figure 1.5.

Neurotrophin signal transduction cascade.

Schematic diagram of intracellular signalling cascade following neurotrophin (NTF) binding to Trk receptors. See text for references and abbreviations.

membrane excitability, that is due to the specific expression of the PN1 sodium channel gene (Toledo-Aral *et al.*, 1995). PN1 gene induction is ras-independent, requires prior synthesis of IEGs and appears 4 hours after the pulse of NGF. In contrast, the continuous exposure to NGF upregulates a larger set of sodium channels, including the brain type II/IIA. Therefore, different genetic programs exist in PC12 cells for the response to NGF signal. PC12 cells and, more generally, NGF-responsive neurons, could use different pathways according to different requirements for trophic support during development.

1.6   Activity-dependent response of neurotrophins in the CNS.

As widely discussed in section 1.1, synaptic plasticity is dramatically dependent on neuronal activity. Therefore, activity-dependent regulation of neurotrophin synthesis and release in the CNS has been proposed to play a key role in trophic interactions that occur at the synapse level during synaptic strengthening and formation. So far, a large series of studies demonstrated that neurotrophins are regulated by neuronal activity. These data have been recently reviewed by Isackson (1995) and will be discussed in this section.

1.6.1   Activity-dependent neurotrophin gene expression.

The first evidence that electrical activity can regulate neurotrophin expression *in vivo* was given by Gall and Isackson (1989), who demonstrated that recurrent limbic seizures increased NGF mRNA in hippocampal neurons. Subsequent studies have demonstrated that neurotrophin mRNAs are up-regulated with specific patterns of expression following seizures induced by electrolytic lesions of the dentate gyrus (Isackson *et al.*, 1991), kindling (Ernfors *et al.*, 1991), or kainic acid (KA) systemic administration (Zafra *et al.*, 1990; Gall *et al.*, 1991; Dugich-Djordjevic *et al.*, 1992; Wetmore *et al.*, 1994).

A large series of studies investigated the role of different neurotransmitter systems in the regulation of seizure-induced neurotrophin expression in hippocampus and neocortex. Glutamatergic and GABAergic systems are presumably involved in this process. Both *in vivo* and *in vitro*, KA up-regulates the levels of NGF and BDNF mRNAs in hippocampal neurons, whereas N-methyl-D-aspartic acid (NMDA) or inhibitors of NMDA glutamate receptors have no effect. Kainic acid-mediated increase of NGF and BDNF mRNAs can be blocked by non-NMDA receptor antagonists (Zafra *et al.*, 1990). In contrast, the blockade of NMDA receptors and/or the stimulation of GABA receptors decreases the basic levels of BDNF and NGF mRNAs in hippocampus (Zafra *et al.*, 1991).

Recent results have also shown that GABAergic stimulation can differentially regulate BDNF mRNA levels in different phases of *in vitro* maturation of hippocampal neurons (Berninger *et al.*, 1995). Thus, whereas non-NMDA receptors regulate basal levels of neurotrophin mRNAs, maintenance of these normal levels seems to be mediated predominantly by GABA and NMDA receptors. Moreover, injections of pilocarpine (an agonist of muscarinic acetylcholine receptors) increase hippocampal BDNF and NGF mRNAs in both early postnatal and adult rats, thus suggesting that also the cholinergic system can positively regulate neurotrophin expression *in vivo* (Berzaghi *et al.*, 1993).

1.6.2   Activity-dependent response of neurotrophins in the visual system.

Results discussed above clearly demonstrate that neurotrophin expression is dependent upon electrical activity. *In vivo* studies are difficult to interpret, as altered neurotrophin response following surgical or pharmacological treatments could result either from increased neuronal activity or, conversely, from the excitotoxic effects of the intense electrical activity. However, some important studies indicate that physiological levels of afferent activity can regulate neurotrophin expression *in vivo*. For example, sub-seizure stimulation sufficient to induce long-term potentiation (LTP) increases neurotrophin mRNA in rat hippocampus (Patterson *et al.*, 1992; Parsadian *et al.*, 1993). Moreover, it has been recently demonstrated that physical exercise can increase BDNF gene expression in hippocampus and neocortex (Neeper *et al.*, 1995).

Interesting results concern the regulation of neurotrophin expression in response to visual stimuli. Castrén *et al.* (1992) have reported that levels of BDNF and TrkB mRNA in the visual cortex are reduced by rearing adult rats in total darkness for 1 week. The exposure to light after this period in darkness restores BDNF and TrkB mRNAs to control levels within 1 hour. In contrast, these treatments do not affect the level of NGF mRNA in the visual cortex. Moreover, intraocular injections of tetrodotoxin (TTX, a sodium channel blocker) in adult rats induce, within 6 hours, a 40% reduction of BDNF mRNA in the visual cortex contralateral to the injected eye. In a more recent paper, Schoups *et al.* (1995) confirmed the effect of dark rearing on BDNF mRNA expression in the visual cortex but, in contrast with Castrén *et al.* (1992), they also observed that dark rearing slightly increased cortical levels of NGF mRNA.

Finally, we have recently shown that monocular deprivation is sufficient to reduce the level of BDNF mRNA in the rat visual cortex (Bozzi *et al.*, 1995). These results will be presented in Chapter 2.

Taken together, these results clearly demonstrate that light deprivation down-regulates the amount of BDNF mRNA in the rat visual cortex, and suggest that different mechanisms could be involved in the modulation of neurotrophin mRNAs by visual experience. So far, these are the most important evidences that physiological levels of sensory inputs (such as light) play a key role in the regulation of neurotrophin gene expression.

1.6.3   Cellular mechanisms regulating neurotrophin gene expression.

So far, few data are available for intracellular mechanisms that regulate activity-dependent neurotrophin gene expression. In cultured cortical neurons, Ghosh *et al.* (1994) found that BDNF acts as a survival factor, and its action is mediated by voltage-sensitive calcium channels (VSCC). Potassium chloride (KCl), but not glutamate, increased survival of cultured embryonic cortical neurons, and this effect was blocked by anti-BDNF antibodies. Both KCl and glutamate increased levels of BDNF mRNA, but KCl-induced BDNF expression was longer lasting and higher than that elicited by glutamate. Moreover, the effect of KCl was mediated by L-type VSCC, while glutamate up-regulated BDNF mRNA through NMDA receptors. Similar results have also been obtained in neuronal cultures of the rat hippocampus, where L-type VSCC blocker nifedipine reduced KCl- and kainic acid-mediated increase in BDNF and NGF mRNAs (Zafra *et al.*, 1992). Voltage gated Na$^+$ channels are also supposed to mediate the activity-dependent expression of NGF, as veratridine (an activator of this type of channels) increases NGF mRNA in hippocampal cultures (Lu *et al.*, 1991).

Multiple transcripts of the BDNF gene have been characterized, and are expressed in different brain regions. The entire pre-pro-BDNF is coded by one exon (exon V) but at least seven different BDNF transcripts containing different 5' exons are generated by differential transcription start-site usage (Timmusk *et al.*, 1993b). A similar genomic structure has also been described for NT-3 (Lindholm, 1994) and is supposed to exist for the NGF gene too. Interestingly, these different BDNF transcripts show distinct pattern of expression following kainic acid-induced seizures (Metsis *et al.*, 1993; Timmusk *et al.*, 1995). Thus, different promoter elements regulate BDNF mRNA expression in response to increased afferent activity. Specific subsets of transcription factors must be involved in the differential regulation of BDNF gene promoters. Analysis of the BDNF gene promoters (Timmusk *et al.*, 1993b, 1995) revealed no consensus sequences for transcription factor AP-1 (TGAGTCA) or zif268 (CGCCCC/GCGC), that are supposed to regulate activity-dependent gene expression in CNS neurons (for a review see Sheng and Greenberg, 1990). So far, the specific factors that bind to BDNF promoters and the signal transduction pathways that lead to their activation are still largely unknown.

Few data are also available for the promoter sequences that drive the expression of other neurotrophins. In the PNS, Hengerer *et al.* (1990) found that a binding site for transcription factor AP-1 (Fos/Jun heterodimer) was present in the NGF gene promoter and increased the expression of NGF mRNA after sciatic nerve lesion. Similar mechanisms could also regulate activity-dependent NGF expression in CNS neurons, but these have not been clearly demonstrated so far.

1.6.4 Activity-dependent synthesis and release of neurotrophin proteins.

In contrast with the large series of studies demonstrating that neurotrophin gene expression is dependent on neuronal activity, only few data are avalaible for activity-dependent neurotrophin release in CNS neurons (reviewed in Thoenen, 1995). In hippocampal neurons, NGF is released in both constitutive and regulated pathways (Blöchl and Thoenen, 1995). Constitutive release is predominantly localized at the surface of the cell body, while high potassium- or glutamate-induced NGF secretion occurs from dendrites. This is consistent with the finding the intracellular stores of NGF are localized both in the pericaryon and neuronal processes, including dendrites (Thoenen, 1995). Activity-dependent secretion of NGF induced by KCl, glutamate or veratridine (an activator of voltage gated Na^+ channels) depends on extracellular Na^+ and intact intracellular calcium stores. Similar results have recently been obtained also for activity-dependent BDNF release from hippocampal neurons (Griesbeck *et al.*, 1995). Thus, neurotrophin release from CNS neurons seems to be regulated by specific mechanisms, different than those used for regulated secretion of neuropeptides and neurotransmitters (that is known to depend on extracellular Ca^{2+}; for a review see Südhof, 1995, and references therein).

Glutamate-induced BDNF synthesis and release from hippocampal neurons involve non-NMDA (kainate) type glutamate receptors (Wetmore *et al.*, 1994). In CA3 hippocampal pyramidal neurons, BDNF immunoreactivity (-IR) is mainly localized in the cytoplasm and dendrites. Two or three hours after kainic acid systemic treatment, BDNF-IR shifts from intacellular compartments to the extracellular matrix, thus suggesting that KA can rapidly induce BDNF local release from the cell body and dendrites of CA3 neurons. BDNF new synthesis is also induced by KA treatment, as cytoplasmic BDNF-IR is increased within 6-8 hours after KA exposure.

Recent results also confirm that BDNF protein synthesis is dependent on afferent electrical activity. By using a novel sensitive two-site immunoassay, Nawa *et al.* (1995) have demonstrated that the amount of BDNF protein markedly increases in the hippocampus and entorhinal cortex/amygdala within two-four hours after the onset of limbic seizures induced by the electrolytic lesion of the dentate gyrus hilus. In limbic structures, BDNF protein levels remains elevated 4 days afer seizure onset. This sustained increase of BDNF protein is paralleled by a prolonged up-regulation of neuropeptide Y in the same structures. Therefore, both short-term and long-term up-regulation of BDNF synthesis and function (e.g. modulation of neuropeptide release; see also the Discussion, for a possible role of BDNF and other neurotrophins in neurotransmitter and neuropeptide secretion) seems to be dependent on increased neuronal activity. Finally, nothing is known about the activity-dependent synthesis and release of BDNF protein in the rat visual system. This problem is currently under investigation in our laboratory.

1.7   Aim of the experimental work.

In this Introduction I have discussed the large series of studies that support the "neurotrophic hypothesis" for the plasticity of the mammalian visual system. In particular, neurotrophins NGF, BDNF and NT4/5 are thought to finely regulate the connectivity in the geniculo-cortical pathway during the critical period. In a recent review, Kevin Fox and Kathleen Zahs defined three major tests for evaluating the possible role of neurotrophic factors in the plasticity of the visual system (Fox and Zahs, 1994):

1) Blocking the action or altering the levels of neurotrophins should affect plasticity.

2) The levels of the neurotrophins (and/or their receptors) should vary with the critical period.

3) Neurotrophin expression should be regulated by neuronal activity, that is known to regulate experience-dependent synaptic plasticity.

Some of the results reported in this Introduction clearly confirm the first two postulates:

1) Anti-NGF antibodies prolong the critical period of plasticity in the rat (Domenici *et al.*, 1994a). Conversely, intracortical injections of both BDNF or NT-4/5 inhibit the formation of ocular dominance columns in the cat (Cabelli *et al.*, 1995a).

2) The amount of both NGF and BDNF mRNAs in the visual cortex varies with the critical period (Castrén *et al.*, 1992; Bozzi *et al.*, 1995; Schoups *et al.*, 1995).

As regards the third postulate proposed by Fox and Zahs, several *in vitro* and *in vivo* studies have demonstrated that neurotrophin production and release are dependent on neuronal activity in the CNS. These results have been already reviewed in this Introduction (see 1.6). However, until few years ago, a direct evidence for an activity-dependent regulation of neurotrophin gene expression in the visual system was still lacking. Monocular deprivation has been largely used by several Authors as an experimental model to study activity-dependent plasticity of the visual system. According to the previously discussed "neurotrophic hypothesis", the effects of monocular deprivation could be ascribed to a reduced production and/or release of an endogenous neurotrophic factor of the NGF family. Four years ago, the work of Castrén and colleagues clearly demonstrated for the first time that light can regulate the cortical levels of BDNF mRNA: rearing young or adult rats in complete darkness decreases the expression of BDNF mRNA in the visual cortex (Castrén *et al.*, 1992). However, the results obtained with dark rearing can not directly apply to monocular deprivation for at least three reasons.

1) During dark rearing there is no competition between the two eyes, in that both eyes are equally deprived.

2) Dark rearing is a rather peculiar manipulation which has the disadvantage of interfering with the normal behavior of the animal, thus not allowing to understand the primary cause of any given effect.

3) Monocular deprivation by eyelids suture deprives the eye of pattern vision, while preserving luminance stimulation.

Therefore, we decided to investigate whether neurotrophins and their receptors could be down-regulated by monocular deprivation either during or after the critical period in the rat. In particular, I focused my attention on the expression of mRNAs for BDNF and TrkB, as these molecules have been reported to be extremely sensitive to light deprivation. The results of these studies will be presented in Chapter 2, that is essentially subdivided in three parts:

1) In the first part (sections 2.1 and 2.2), I will show that monocular deprivation can down-regulate the expression of BDNF (but not TrkB) mRNA in the rat visual cortex. I will also report a detailed analysis of the temporal and spatial distribution of BDNF and TrkB mRNAs in the visual cortex and dLGN during the critical period. Preliminary results on the locatization of BDNF and Trk proteins in the visual cortex have been included.

2) To further investigate the electrical activity-dependance of BDNF mRNA expression in the visual cortex, I studied the effects of two different pharmacological treatments: intraocular injections of tetrodotoxin and intracortical bicuculline administrations. These protocols have been used, respectively, to block or increase the electrical activity of visual cortical neurons (section 2.3).

3) In the last part (section 2.4), I will analyse the response of the rat visual cortex to neurotrophins during the critical period. So far, contrasting results have been reported for the action of neurotrophins (NGF, BDNF and NT-4/5) on visual cortical neurons (see above, in this Introduction). Moreover, the complete knowledge of the intracellular mode of action of neurotrophins *in vivo* is still lacking. Therefore, we used a sensitive assay to investigate Trk/neurotrophin receptor activity. In these experiments, NGF and BDNF were locally administered onto the visual cortex and ligand-induced Trk-tyrosine phosphorylation was measured. Preliminary results suggest that Trk/neurotrophin receptors are constantly activated in the visual cortex.

2. RESULTS

2.1   Distribution of BDNF and its receptor (TrkB) in the rat visual system.

2.1.1   Developmental expression of BDNF mRNA in the visual cortex.

The developmental expression of BDNF mRNA in the rat visual cortex was analysed by using quantitative RNAase protection assays. BDNF mRNA is expressed in the rat visual cortex during the whole postnatal life. As shown in Table I, BDNF mRNA is already expressed in the visual cortex at postnatal day (P) 5. Its level of expression rapidly increases around the time of natural eye-opening (P15, beginning of the critical period), and remains at a plateau level from P20 to adult age. Table II reports the quantification of three different experiments, obtained by densitometric analysis. A 10-fold increase in the level of BDNF mRNA is present between P10 and P20.

The spatial distribution of BDNF mRNA in the rat visual cortex was investigated by *in situ* hybridisation. BDNF mRNA is predominantly expressed in cortical layers II-III and V-VI. This pattern of expression was observed at all ages, both during the critical period (P23, P30, P45) and in adult animals (P60) and also in nonvisual cortices. Table III illustrates the labelling of BDNF mRNA in the visual cortex at postnatal days 23, 45 and 60. The typical pattern of expression of BDNF mRNA in cortical layers II-III and V-VI is also shown in Tables VII (A) and VIII (A), that are taken from the visual cortex of a thirty days old rat. BDNF mRNA labelling is not restricted to the visual cortex. Table VI (A) shows a P45 brain slice hybridised with the BDNF probe. All the neocortex and other brain structures are clearly labelled (pontine nuclei, Pn, and hippocampal dentate gyrus, DG). This observation is consistent with previous data from other Authors (Hofer *et al.*, 1990; Phillips *et al.*, 1990; Isackson *et al.*, 1991; Castrén *et al.*, 1992).

Bright field-images taken at higher magnification (Table IX, A) show that BDNF mRNA labelling in the visual cortex is mainly restricted to neurons, as demonstrated by their relatively large cell body and large nucleus. However, the specific subset of cells expressing BDNF remains to be clearly determined.

2.1.2   Developmental expression of TrkB mRNA in dLGN and visual cortex.

The developmental expression of mRNA for BDNF high affinity receptor (TrkB) in the rat visual system was first analysed by RNAase protection. A specific probe for the intracellular tyrosine-kinase domain of TrkB (TrkB-TK$^+$) was used to detect the full-length isoform of TrkB transcript. TrkB mRNA is present in the rat visual cortex during the whole postnatal life. As shown in Table IV, TrkB mRNA is already expressed in the visual cortex at P5, its level of expression remains constant until P20 and then decreases (~70%) until the adult age.

In situ hybridisation experiments were also performed to investigate TrkB mRNA expression in the visual system. Both dLGN and visual cortex express TrkB mRNA at all postnatal ages between P10 and P60, as shown in Tables V and VII (B and C). In dLGN, TrkB is already present at P10, reaches a maximum level around P20-30 and then decreases until adult age (Table V A'-D'). In the visual cortex, TrkB mRNA is uniformly distributed through all cortical layers at all postnatal ages. TrkB expression is higher at P23 and then slightly decreases until P60 (Table V A-D). Tables VII (B, C) and VIII (B) show the typical pattern of expression of TrkB mRNA in the visual cortex and dLGN at P30. Table VI (B) shows that the TrkB transcript is widely distributed throughout the rat brain. Neocortex (ctx), hippocampus (hip) and thalamic nuclei other than dLGN are strongly labelled by the TrkB-TK$^+$ probe. In the visual cortex, TrkB mRNA labelling is mainly restricted to neurons, as shown in Table IX (B). Preliminary results from other Authors suggest that TrkB mRNA is mainly localized in both pyramidal neurons and GABAergic (parvalbumine-positive) interneurons (Cellerino *et al.*, 1995). However, also few glial cells seem to be weakly labelled by the TrkB-TK$^+$ probe (Table IX, B).

2.1.3   BDNF immunoreactivity in the visual cortex.

The distribution of BDNF protein in the visual cortex was investigated by immunohystochemistry with a polyclonal antibody preparation made against recombinant-human BDNF (rhBDNF). The specificity of the antibody was tested by immunoblot. The anti-BDNF antiserum recognized only purified rhBDNF whereas no cross-reactivity was detected with either mouse NGF, human NT-3 or human NT-4/5 (data not shown).

Coronal sections of visual cortex of 30 days-old rats were probed with the anti-BDNF antiserum. A strong labelling was found in the deep layers V-VI. In the superficial layers II-III, the labelling was weaker, and only few neurons in layer IV showed BDNF immunoreactivity (-IR) (Table X). Both pyramidal and non-pyramidal neurons were labelled. In all positively stained cells, BDNF-IR was confined to the cytoplasmic area and dendrites, but not to the nuclei (Table X, B and C). Glial cells were not labelled. Therefore, the distribution of BDNF protein in the rat visual cortex seems to be comparable to that of BDNF mRNA. In a series of control experiments, the anti-BDNF antibody recognized exogenous BDNF (3-5 μg) injected into the cortex of rats (data not shown). Moreover, preabsorption with rhBDNF completely abolished the immunoreactivity in brain sections (data not shown).

2.1.4   Trk-like immunoreactivity in the visual cortex.

The developmental expression of Trks-neurotrophin receptors in the rat visual cortex was analysed by immunoblot with an anti-panTrk polyclonal antibody. This antibody recognizes 14 aminoacids shared by the carboxy-terminal of TrkA, TrkB and TrkC. Results are shown in Table XI. Trk-like immunoreactivity in the visual cortex is already present at P10 and remains constant until P45. In the same experiment, basal forebrain was used as a positive control, since it is known to express TrkA, TrkB and TrkC (see Introduction). Trks are present in the basal forebrain at P10 and their expression decreases until P30, according to previous findings from other Authors (Li *et al.*, 1995) (Table XI). These results clearly show that during the critical period the rat visual cortex contains the Trk receptors and can therefore be responsive to the action of different neurotrophic factors.

2.2   Visual experience regulates BDNF (but not TrkB) mRNA in the rat visual cortex.

2.2.1   Monocular deprivation during the critical period.

Monocular deprivation represents a classical model for the study of plasticity in the visual system. The "neurotrophic hypothesis" previously described (see Introduction) suggests that the anatomical and physiological effects of monocular deprivation could be ascribed to a reduced synthesis of endogenous neurotrophin(s), determined by an altered (non-physiological) electrical input to the visual cortex. To test this hypothesis, BDNF mRNA expression was analysed by *in situ* hybridisation in P45 rats monocularly deprived from P15 (this deprivation covers the entire critical period of the rat; Fagiolini *et al.*, 1994a). The level of BDNF mRNA was remarkably reduced in the visual cortex contralateral to the deprived eye (Table XII). The presence of the effect of MD only in the cortex contralateral to the deprived eye is not surprising, considering the organisation of the rat visual pathway. The reduction of BDNF mRNA labelling induced by MD equally affected primary and secondary visual areas (indicated by arrows, in Table XII, B) and was not present in nonvisual areas. MD appeared less effective in the binocular subfield of the primary visual cortex (area Oc1B, Table XII, B). This observation can be explained by the presence of an active input from the nondeprived eye. As shown in Table XII (D), the reduction of BDNF mRNA is more apparent in superficial layers II/III. Slices of visual cortex of P45 nondeprived rats hybridised on the same slides were used as control and did not reveal any difference in BDNF mRNA expression between left and right cortices (Table XII, E). Two weeks of MD, from P15 to P30, gave comparable results (data not shown). RNAase protection confirmed the results obtained by *in situ* hybridisation. In rats monocularly deprived from P15 until P30, BDNF mRNA level was reduced in the visual cortex contralateral to the deprived eye, as compared to the ipsilateral cortex (Table XIII). Levels of β-actin mRNA, used as an internal standard for RNA quantification, were not affected by MD. Densitometric analysis of the autoradiograms revealed a 55% decrease of BDNF mRNA amount in the cortex contralateral to the deprived eye.

2.2.2   *Monocular deprivation in adult rats.*

The classical effects of MD on the visual system are restricted to a critical period of postnatal development (Shatz and Stryker, 1978; Fox and Zahs, 1994) and do not occur when MD is performed in adults. For these reason, we sought to determine whether MD in adult animals could decrease the level of BDNF mRNA in the visual cortex. Rats were monocularly deprived after the end of the critical period, from P45 to P60. At the end of the deprivation period, BDNF mRNA expression was analysed by *in situ* hybridisation.

Levels of BDNF mRNA were remarkably reduced in the visual cortex contralateral to the deprived eye (Table XIV). No difference between left and right cortices was observed in slices obtained from P60 nondeprived animals hybridised on the same slide (data not shown). As indicated in Table XIV (A), BDNF mRNA only decreased in the primary and secondary visual areas (between arrows) and not in pontine nuclei (Pn) or other nonvisual areas. As already observed in MD young rats, the reduction of BDNF mRNA is less evident in the binocular subfield (area Oc1B, Table XIV, C). It is likely that the presence of an active input from the nondeprived eye can sustain BDNF mRNA expression in area Oc1B. BDNF mRNA reduction is more evident in the superficial layers II-III (Table XIV, B and C).

2.2.3 *Monocular deprivation does not affect TrkB mRNA expression in dLGN and visual cortex.*

The "neurotrophic hypothesis" for the plasticity of the visual system postulates that electrical activity could regulate the production, release and uptake of neurotrophic factors in the visual cortex. Results described above, together with previous findings from other Authors (Castrén *et al.*, 1992; Schoups *et al.*, 1995) clearly demonstrate that light regulates neurotrophin mRNA expression in the visual cortex. Recent observations (reviewed in Thoenen, 1995) also suggest that neurotrophin release could be activity-dependent. Conversely, no data are available about the regulation of neurotrophin uptake. Electrical activity could regulate neurotrophin uptake by altering the production of their specific receptors. To test this hypothesis, we investigated whether monocular deprivation affected the levels of mRNA for the high affinity BDNF receptor (TrkB) in the visual system. Sections obtained from the same monocularly deprived rats adopted for BDNF mRNA analysis were processed for *in situ* hybridisation with a TrkB-TK+ riboprobe.

Monocular deprivation, both during the critical period and in adult animals, did not alter TrkB mRNA expression. In rats monocularly deprived from P15 to P30 or from P15 to P45, TrkB mRNA remained uniformly dstributed through all the layers of both left and right visual cortices (Table XV, A). In the same animals, no difference in TrkB mRNA levels was observed in the dLGN ipsilateral and contralateral to the deprived eye (Table XV, B and C). Same results were obtained when MD was performed for two weeks in adult animals (data not shown).

2.3 Pharmacological regulation of BDNF mRNA in the visual cortex.

The electrical activity-dependence of BDNF mRNA expression in the rat visual cortex was further investigated by different pharmacological approaches. In particular, activity of visual cortical neurons was either down- and up-regulated by two different treatments: intraocular injections of tetrodotoxin and local administration of bicuculline.

2.3.1 *Intraocular injections of tetrodotoxin.*

Intraocular injections of tetrodotoxin (TTX, a sodium channel blocker) were used to block the afferent input to the visual cortex. In particular, the effects of pharmacological monocular blockade of retinal spike activity were compared during or after the critical period, to further quantify the age-dependence of the regulation of BDNF mRNA by visual input. By using RNAase Protection experiments, we demonstrated that intraocular TTX injections reduced BDNF mRNA in the contralateral visual cortex with comparable efficacy in young and adult rats (Table XVI). In agreement with Castrén *et al.* (1992), TTX treatment in adult rats determined a reduction of 47.9±7.9% (n=3) of BDNF mRNA. The effect of TTX in rats within the critical period was not significantly different (36±2.6 %, n=3. p>0.05, Student's t-test).

2.3.2 *Intracortical administration of bicuculline.*

In a different experimental protocol, we determined whether BDNF mRNA expression could be up-regulated by increasing electrical activity of the visual cortex. Bicuculline (a $GABA_A$ receptor antagonist) was applied onto the visual cortex to determine a local post-synaptic activation of visual cortical neurons. The effects of this treatment were controlled electrophysiologically in parallel experiments. Spike activity of single neurons of the primary visual cortex was recorded. Recordings were performed from the cortex treated with bicuculline and from the contralateral cortex treated with vehicle. We found that in 15 min bicuculline treatment dramatically affected the cell discharge inducing high frequency (approximately 100 Hz) train of spikes (data not shown). Each train appeared every 1.5-2 s and lasted for 300-400 ms. This pattern of activity was observed for at least 8 h. The change in activity was present only in the bicuculline treated cortex.

RNAase protection revealed that the level of BDNF mRNA was markedly increased in the visual cortex treated with bicuculline, compared with the contralateral cortex treated with vehicle (Table XVII, A). Levels of β-actin mRNA, used as an internal standard for RNA quantification, remained unaltered both in left and right cortices. Densitometric analysis of the autoradiograms revealed a five-fold increase of BDNF mRNA expression. *In situ* hybridisation performed on P30 rats treated with bicuculline revealed that BDNF mRNA increase in the treated cortex was distributed throughout all the layers. However, the effect was more pronounced on layers II-III (Table XVII, B).

2.4   The mechanism of action of neurotrophic factors in the visual cortex: tyrosine phosphorylation of Trk-like receptors.

To further analyse the *in vivo* mechanism of action of neurotrophins, NGF- and BDNF-induced tyrosine phosphorylation of Trk proteins was studied in the rat visual cortex.

2.4.1   *Basal and neurotrophin-induced Trk-tyrosine phosphorylation in the visual cortex.*

To elucidate the mechanism of action of neurotrophic factors in the visual cortex, it is important to understand how they can activate Trk receptors *in vivo*. The phosphorylation of specific tyrosine residues in the intracellular domain of Trk receptors is the first step of their activation in response to neurotrophin binding. For these reasons, I investigated whether NGF and BDNF could induce Trk-tyrosine phosphorylation in the rat visual cortex during the critical period. The experimental protocol was set up by a series of preliminary *in vitro* experiments. Visual cortices and basal forebrains were dissected from P20-21 rats and treated *in vitro* with NGF and BDNF (see legend of Table XVIII for details). Homogenised tissues were immunoprecipitated with anti-panTrk antibody and immunoblotted with anti-phosphotyrosine or anti-panTrk antibodies. The result of these experiments is shown in Table XVIII (A). In the basal forebrain, both NGF and BDNF induced tyrosine phosphorylation of Trk-like proteins, while no signal was present in control samples treated with Krebs' medium alone. Interestingly, in the visual cortex, Trk-tyrosine phosphorylation was present both in untreated and

NGF- or BDNF- treated samples. To quantify Trk-like proteins, the same filter was washed and reprobed with the anti-panTrk antibody. All samples showed a comparable signal (Table XVIII, B).

These results were confirmed by *in vivo* neurotrophin treatment. NGF and BDNF were locally applied onto the right visual cortex of 23-24 day-old rats. Thirty minutes after, visual cortices were analysed for Trk-tyrosine phosphorylation. Results are shown in Table XIX (A). As already observed for *in vitro* NGF-treatment, Trk-tyrosine phosphorylation was both present in the visual cortex treated with NGF or vehicle (saline) solution. A comparable level of Trk-tyrosine phosphorylation was revealed in untreated frontal cortices dissected from the same animal. The same results were obtained for the BDNF treatment. No increase in Trk-tyrosine phosphorylation was present in the BDNF treated visual cortex, as compared with saline treated visual cortex or untreated frontal cortices. The level of Trk-like receptors was comparable in all samples, as revealed by reprobing the same filter with the anti-panTrk antibody (Table XIX, B). These data suggest that a basal activity of neurotrophin receptors is present in the rat visual cortex during the critical period.

3. DISCUSSION

3.1   Brief summary of results.

In the previous Chapter I reported the results of my study of the developmental expression of BDNF mRNA in the rat visual cortex and its regulation by monocular deprivation. Monocular deprivation of pattern vision by eyelids suture reduces the cortical amount of BDNF mRNA during the critical period. This treatment has the same effects in the adult animal, where this procedure does not induce anatomical or physiological deficits. Conversely, MD does not affect the amount of mRNA for TrkB/BDNF receptor. The same results were also obtained by intraocular injections of TTX, that completely silence retinal spike activity. Moreover, intracortical administration of bicuculline (an antagonist of GABA$_A$ receptor) markedly increases the amount of BDNF mRNA. BDNF mRNA and protein are not uniformely distributed in the visual cortex during the critical period and adulthood. Very low levels of labelling in layer IV suggest that this neurotrophin does not act directly on thalamic afferents. The uniform distribution of TrkB mRNA throughout all cortical layers are in favor of a possible intracortical action of BDNF. Finally, preliminary results show that a basal activity of neurotrophin Trk receptors is present in the rat visual cortex during the critical period, thus suggesting a constitutive action of neurotrophins during visual cortical development. These results strongly confirm a possible involvement of BDNF (and, more generally, of neurotrophins) in the molecular mechanisms that regulate both the establishment and maintenance of visual cortical functions. Here I discuss the recent findings from other Authors, that support this hypothesis.

3.2   Developmental expression of BDNF and TrkB in the rat visual system.

3.2.1   *BDNF mRNA in the rat visual cortex increases at the time of eye opening.*

The development of the visual cortex involves activity-dependent refinement of axonal arborizations and synapse formation and maturation. Specific factors regulated during development are likely to participate in this process. Our results support the hypothesis that the product of the BDNF gene could be one of these factors. Data show that BDNF mRNA is present in the visual cortex during the rat critical period, it increases ten fold between P10 and P20, and after P20 no significant change is observed. A possible interpretation of this result (considering that the time of natural eye opening in rats is at P14-15) is that eye opening could up-regulate BDNF mRNA expression by inducing a sudden increase in the electrical activity in the afferents to the visual cortex. It is interesting to note that these results are in complete agreement with those obtained by Castrén *et al.* (1992) and, more recently, also by Schoups *et al.* (1995).

3.2.2   *Distribution of BDNF mRNA and protein in the visual cortex.*

The spatial pattern of BDNF mRNA labelling revealed by *in situ* hybridisation shows an uneven distribution throughout the different layers of the visual cortex. BDNF mRNA is high in layers II-III and V-VI but very low in layer IV. This pattern of distribution is present at all the ages and was observed also in nonvisual cortices (Hofer *et al.*, 1990; Phillips *et al.*, 1990; Isackson *et al.*, 1991). Interestingly, BDNF immunoreactivity (-IR) in the visual cortex shows a comparable localisation to that of BDNF mRNA. A clear labelling is present in dendrites and cell bodies of both pyramidal and non-pyramidal neurons in the superficial layers II-III and deep layers V-VI, whereas only few neurons in layer IV are BDNF-IR positive. These results are in complete agreement with those obtained by other Authors in the cat and ferret visual cortex (Cabelli *et al.*, 1995b). The specific subset of cells expressing BDNF and TrkB in the visual system remains to be clearly determined. It has been suggested that BDNF protein could be localised in parvalbumin-positive GABAergic neurons of the neocortex (Berzaghi *et al.*, 1994). However, more recent data argue against this observation and suggest that in the adult rat visual cortex BDNF protein is mainly localised in non-GABAergic pyramidal neurons (Cellerino *et al.*, 1995).

3.2.3   *Distribution of TrkB in dLGN and visual cortex.*

In situ hybridisation experiments reveal that mRNA for the full-length isoform of TrkB/BDNF high affinity receptor is widely expressed in the visual system during all postnatal development. In dLGN, TrkB is already present between P5 and P10, reaches a maximum level around P20-30 and then decreases until adult age. This result is in good agreement with previous observations from other Authors (Ringstedt *et al.*, 1993; Schoups *et al.*, 1995). In the visual cortex, TrkB mRNA is uniformly distributed through all cortical layers at all postnatal ages. TrkB mRNA expression is higher at P23 and then slightly decreases until P60. TrkB mRNA labelling in the visual cortex is mainly restricted to neurons, but a TrkB labelling also on glial cells (that are known to express TrkB mRNA; Frisén *et al.*, 1995) can not be completely excluded. However, recent results strongly suggest that TrkB protein is localised in both pyramidal and non-pyramidal neurons of the adult rat visual cortex (Cellerino *et al.*, 1995). In particular, most of non-pyramidal neurons labelled by anti-TrkB antibodies are also parvalbumin (PV) positive, thus suggesting that GABAergic, PV expressing interneurons also express TrkB protein. This is supported by the observation that PV stained cells also contain TrkB mRNA. Conversely, pyramidal neurons positive for TrkB immunoreactivity are surrounded by PV-containing synaptic boutons.

As regards the developmental expression of TrkB protein in the rat visual cortex, the only data that I presented concerns an immunoblot analysis of Trk-like proteins obtained with an anti-panTrk polyclonal antibody. This antibody recognizes all the members of the Trk family of neurotrophin recptors (TrkA, TrkB and TrkC). Trk-like immunoreactivity (-IR) in the visual cortex is already present at P10 and remains constant until P45. Results obtained by other Authors strongly suggest that TrkB protein is the more abundant among the Trk receptors in the visual cortex (see Allendoerfer *et al.*, 1994). Therefore, the pattern of panTrk-IR could mainly reflect that of TrkB expression. However, immunoblot analysis with anti-TrkB antibodies should be done to confirm this hypothesis.

3.3   The action of BDNF in the visual cortex.

The action of BDNF on target cells is mediated by its specific high-affinity receptor TrkB (Klein *et al.*, 1989, 1991b). Several truncated isoforms of this protein have been discovered. Truncated isoforms lack the intracellular domain of the protein that bears the tyrosine-kinase activity and seem unable to transduce the signal (Klein *et al.*, 1990; Middlemas *et al.*, 1991). In order to understand the possible site of action of BDNF I therefore limited my study to the full-length isoform of TrkB, that seems to be responsible for BDNF signal transduction. Data show that the mRNA for the full length isoform of TrkB, that contains the tyrosine-kinase domain, is present in the rat dLGN during the critical period. There is evidence that cells in dLGN express TrkB protein: iodinated BDNF can be cross-linked to full length TrkB in the ferret dLGN during the period of ocular dominance column formation (Allendoerfer *et al.*, 1994). These observations suggest that geniculate neurons are able to transduce BDNF signalling during the critical period. However, as in other mammals, rat thalamic innervation to the visual cortex is restricted to layer IV with a small input to layer VI (Kageyama and Robertson, 1993). The weak BDNF mRNA labelling in layer IV argues against an action of BDNF on thalamic fibres. Cortical BDNF could act on systems other than dLGN. At least three different hypotheses seem likely.

1) BDNF could act on the cholinergic neurons of the basal forebrain. These neurons are responsive to neurotrophins *in vivo* (Knüsel *et al.*, 1992) and *in vitro* (Alderson *et al.*, 1990) and send projections to the visual cortex (Carey and Rieck, 1987). It has been shown that cholinergic neurons of the basal forebrain can retrogradely transport radioactively labelled BDNF (DiStefano *et al.*, 1992).

2) BDNF could act, in a paracrine or autocrine mode of action (Korsching, 1993) on visual cortical cells expressing TrkB. TrkB mRNA is present in all layers of the visual cortex. In particular, recent data suggest that BDNF synthesized by pyramidal neurons (Wetmore *et al.*, 1990) could act on GABAergic interneurons in the adult rat visual cortex (Cellerino *et al.*, 1995)

3) Finally, BDNF could act on cortical glial cells, that are known to express TrkB mRNA (Frisén *et al.*, 1995) and have been found to respond to BDNF stimulation *in vitro*. In rat cortical glial cells in culture, BDNF induces a rapid increase in tyrosine phosphorylation of MAP kinase (ERK1-ERK2), MAP kinase activity, intracellular Ca^{2+} concentration and c-fos mRNA expression (Roback *et al.*, 1995). This results clearly show that BDNF can act also on glia, by activating a specific neurotrophin signal transduction pathway in these cells.

3.4   Activity-dependent BDNF expression in the rat visual cortex.

3.4.1   *Light regulates gene expression in the mammalian visual cortex.*

Visual   experience during the critical period strongly affects the physiology and anatomy of the mammalian visual cortex. These effects are likely to be mediated by specific genes whose expression in cortical neurons is finely regulated by small differences in the quantity and pattern of afferent electrical activity. A large series of studies over the past ten years have demonstrated that visual experience can regulate gene and protein expression in the mammalian striate cortex. Thus, "epigenetic" factors can strongly modify the physiology of the visual cortex also at the cellular and molecular level. For example, Hendry and Kennedy have demonstrated that monocular deprivation can selectively increase the immunoreactivity for type II Ca^{2+}/calmodulin kinase (CaM kinase II) in dominance columns connected with the deprived eye in the monkey area 17 (Hendry and Kennedy, 1986). Conversely, monocular deprivation reduces the number of GABA-positively stained cells in deprived-eye dominance columns (Hendry and Jones, 1986). Interestingly, other Authors showed that both CaM kinase II and glutamic acid decarboxylase (GAD, the GABA synthetic enzyme) are markedly increased by rearing kittens in complete darkness for one week (Neve and Bear, 1989). In more recent years, the expression of *zif268* transcription factor mRNA and protein in primary visual cortex has been found to be extremely sensitive to light deprivation in rats and monkeys (Worley *et al.*, 1991; Chaudhuri and Cynader, 1993). Finally, a set of genes expressed in response to light in the adult rat visual cortex (CPGs, cortical plasticity genes) has been recently described (Nedivi *et al.*, 1996). Thus, cortical neurons respond to visual experience by finely regulating specific sets of genes and proteins presumably involved in the early and/or late cellular response to the stimulus (such as kinases, transcription factors and neurotransmitter biosynthetic enzymes).

3.4.2   *Monocular deprivation decreases BDNF mRNA in the rat visual cortex.*

The results outlined above clearly suggest that a large series of different genes mediate the physiological process of segregation of geniculo-cortical connections in the mammalian visual cortex. BDNF could be one of these mediators, since its expression can be down-regulated by monocular deprivation in the rat visual cortex. BDNF receptor TrkB could be also involved in this process. By using the Northern blot technique, a small reduction of TrkB mRNA in the visual cortex of dark-reared rats has been observed (Castrén *et al.*, 1992). My results, obtained by *in situ* hybridisation, show that monocular deprivation does not affect TrkB mRNA expression in the rat visual cortex. A possible interpretation of these data is that modulation of TrkB mRNA expression is not involved in the molecular mechanisms of monocular deprivation. There is still the possibility that small changes in TrkB mRNA are not detected by *in situ* hybridisation.

Monocular deprivation does not arrest the electrical activity of neurons of the visual cortex. The deprived pathway is still spontaneously active and luminance variations are detected and translated into electrical activity by the deprived eye. Moreover, in the binocular subfield (area Oc1B) the great majority of cortical neurons are binocular (80% in rats) (Maffei *et al.*, 1992) and monocular deprivation affects only one of their inputs. Indeed, the effects of monocular deprivation in adult animals are practically absent both at anatomical and physiological levels.

BDNF mRNA down-regulation caused by MD is probably not sufficient to explain the effects of monocular deprivation on the rat visual system. BDNF mRNA labelling is reduced with similar efficacy before or after the end of the critical period and the reduction is also present in the monocular subfield of the primary visual cortex where monocular deprivation is ineffective (Sherman and Spear, 1982). Since the reduction of BDNF mRNA is comparable in young and adult MD animals in the binocular and monocular subfields, this decrease could be plausibly related to the absence of pattern vision, rather than to binocular competition that is effective only during the critical period. Moreover, dark rearing, which does not involve binocular competition, has been reported to affect cortical BDNF mRNA levels similarly to MD (Castrén *et al.*, 1992).

3.4.3   *Cellular mechanisms that regulate activity-dependent BDNF mRNA expression in the rat visual cortex.*

As mentioned above, results obtained with MD in young and adult animals suggest that the pattern of afferent retinal activity is an important factor in the regulation of BDNF mRNA expression in visual cortical neurons. This is consistent with the observation that monocular blockade of retinal spike activity by intraocular injections of the sodium channel blocker TTX produce a ~40% reduction of BDNF mRNA in the visual cortex contralateral to the injected eye. Interestingly, TTX injections have the same effect both during or after the critical period, thus suggesting that the same activity-dependent cellular mechanisms are responsible for BDNF regulation during different phases of cortical development. Moreover, a basal (activity-independent) level of BDNF transcription must exist in the visual cortex, since TTX treatment does not completely abolish BDNF mRNA expression.

As regards the molecular mechanisms that regulate BDNF transcription in response to physiological visual stimuli, I can only suggest some simple considerations. First, we do not know if the BDNF down-regulation induced by TTX (or, conversely, the up-regulation induced by bicuculline) is due to an altered transcriptional rate or to an altered stability of the BDNF messenger RNA. However, recent results from other studies suggest a possible answer to this question. Castrén et al. have demonstrated that kainic acid treatment up-regulates BDNF transcription in the hippocampus without influencing the stability of its mRNA (Castrén et al., 1995). Thus, afferent electrical activity could act on the rate of transcription of the BDNF gene.

A second point concerns the role of different neurotransmitter systems in the basal and activity-dependent expression of BDNF transcript in vivo. As suggested by the large series of pharmacological studies reviewed in the Introduction, acetylcholine, glutamate and GABA all play a critical role in this process. Results obtained with intracortical administration of bicuculline clearly confirm the involvement of a GABAergic inhibitory input in the basal regulation of BDNF expression in the rat neocortex. Removal of this input dramatically increases the amount of BDNF mRNA. Interestingly, electrophysiological recordings were also performed in our experiments to evaluate the effects of bicuculline treatment. Neuron spike activity was extracellularly recorded from the cortex treated with bicuculline and from the contralateral cortex treated with vehicle solution. It was found that in 15 min, bicuculline treatment dramatically affected the cell discharge inducing high frequency train of spikes. This change in activity was present only in the bicuculline treated cortex and no epileptic seizure occurred after the local administration of bicuculline. Therefore, bicuculline-induced BDNF up-regulation is presumably related to increased neuronal activity and not to the excitotoxic effects of the stress-induced response of cortical neurons following bicuculline treatment.

As mentioned above, the maintenance of normal levels of BDNF expression is supposed to be mediated predominantly by GABA and NMDA receptors (NMDAR). We showed that BDNF mRNA expression in the visual cortex increases at the time of natural eye-opening (postnatal day 14-15) and remains at a plateau level for the whole critical period. This developmental regulation could be sustained by the activity of NMDAR. First, at postnatal day 21 NMDAR are mainly expressed in layer II-III of the neocortex (reviewed in Fox and Zahs, 1994), that also express BDNF. Moreover, in the rat visual cortex, the kinetics (Carmignoto and Vicini, 1992) and distribution (reviewed in Fox and Zahs, 1994) of NMDAR shows a good correlation with the critical period for ocular dominance plasticity (Fagiolini *et al.*, 1994a; Fox, 1995), that has been recently shown to overlap, at least in part, with the critical period for long-term potentiation in the visual cortex (Kirkwood *et al.*, 1995). Therefore, several clues suggest that NMDAR can play a critical role in the regulation of BDNF mRNA expression in the rat visual cortex, even though a direct evidence for this hypothesis is still lacking.

Another interesting point concerns the time course of activity-dependent BDNF expression. In my experiments, I did not address this issue. However, several findings support the hypothesis that BDNF is regulated as an immediate early gene in hippocampus and neocortex. For example, kainate-induced increase of BDNF mRNA is not blocked by the inhibitor of protein synthesis cycloheximide (Castrén *et al.*, 1995) and rapidly occurs (within 1-2 hours) after the onset of seizures (see Nawa *et al.*, 1995, and references therein). Finally, Castrén *et al.* (1992) have also demonstrated that BDNF mRNA is rapidly up-regulated by physiological stimuli: the amount of BDNF in the visual cortex is restored to control levels when dark-reared rats are exposed for 1 hour to light. However, the intracellular signal transduction pathways that regulate the basal and activity-dependent expression of the BDNF gene are still largely unknown.

3.5   The mechanism of action of neurotrophins in the rat visual cortex: constitutive activation of Trk-like receptors?

Neurotrophic factors exert their *in vivo* action through the binding to tyrosine-kinase receptors of the Trk family. Trans-autophosphorylation in specific tyrosine residues is the first step of neurotrophin-induced Trk activation (see Introduction). To further analyse the *in vivo* mechanism of action of neurotrophins, the expression of Trk proteins in the rat visual cortex and their activation by both NGF and BDNF have been investigated in this study.

3.5.1   Trk-like immunoreactivity is present in the visual cortex.

Trk-like immunoreactivity in the visual cortex is already present at P10 and remains constant until P45. The anti-panTrk antibody that I used recognizes all the members of the Trk family. From a large series of studies it is well known that the visual cortex express both TrkA and TrkB proteins. TrkA receptors are mainly present on the cholinergic afferents from the basal forebrain (see references cited in the Introduction), whereas TrkB is expressed both on synaptic terminals from geniculate nucleus (Allendoerfer *et al.*, 1994) and GABAergic neurons (Cellerino *et al.*, 1995). Therefore, during the critical period the rat visual cortex contains the Trk receptors and can therefore be responsive to the action of different neurotrophins (NGF, BDNF, and NT-4/5).

3.5.2   Basal Trk-tyrosine phosphorylation in the visual cortex.

To elucidate the mechanism of action of neurotrophic factors in the visual cortex, we investigated whether NGF and BDNF could induce Trk-tyrosine phosphorylation in the rat visual cortex during the critical period. In a series of preliminary experiments, neither NGF nor BDNF could induce Trk-tyrosine phosphorylation in the visual cortex of P21-24 rats both *in vitro* and *in vivo*. Conversely, in the basal forebrain *in vitro*, both NGF and BDNF induced tyrosine phosphorylation of Trk-like proteins, while no signal was present in control samples. These data suggest that a basal activity of neurotrophin receptors is present in the rat visual cortex during the critical period and is presumably due to the action of different neurotrophins released by cortical neurons.

These results are consistent with the observation that NGF treatment can not increase the activity of Choline Acetyl Transferase (ChAT, the synthetic enzyme for acetylcholine) in the rat visual cortex (Domenici *et al.*, 1991). Thus, also the activity of cholinergic fibres in the visual cortex would be maintained to a basal level by different neurotrophins (NGF, BDNF) produced by cortical neurons.

Preliminary results from the group of D. Lindholm (Max Planck Institute für Psychiatrie, Martinsried, Germany) strongly suggest that the basal activation of Trk receptors in the rat cortex could be mediated by glial cells. Primary cultures of astrocytes from the rat cortex express both TrkA mRNA and protein (Dan Lindholm, personal communication). Interestingly, TrkA receptor is constantly phosphorylated on tyrosine residues in these cells, and this basal level of tyrosine phosphorylation does not seem to be further increased by NGF treatment (Dan Lindholm, personal communication).

In our experiments, the lack of neurotrophin-induced Trk phosphorylation in the visual cortex is not certainly due to a short duration (and so, to a reduced efficacy) of neurotrophin treatment. Both *in vitro* (5 min) and *in vivo* (30 min), the duration of the treatment was absolutely comparable to that previously adopted by other Authors in similar experiments (Knüsel *et al.*, 1994; Li *et al.*, 1995). Moreover, both NGF and BDNF were able to induce Trk phosphorylation in the basal forebrain *in vitro*. Therefore, it reasonable to assume that neurotrophin treatment *per se* should be completely effective.

Previous results from other Authors also demonstrate that a basal level of Trk phosphorylation is present in the rat cortex, and that neurotrophin-induced Trk phosphorylation is hard to detect in this brain area at the same postnatal ages as those I analysed (P20-24) (Knüsel *et al.*, 1994). These results could be explained by postulating that Trk receptors could be desensitized *in vivo* by prolonged exposures to neurotrophins. Recent results are consistent with this hypothesis. In particular, it has been demonstrated that intracerebral injection of NGF results in a sustained tyrosine phosphorylation of Trk receptors in the striatum (Li *et al.*, 1995; Peterson *et al.*, 1995). Trk tyrosine phosphorylation is maximal up to 3 days after injection, and is not increased by further exposure to NGF (Peterson *et al.*, 1995). In rat cerebellar granule cells, prolonged exposure to BDNF results in a strong reduction of BDNF binding to TrkB. This reduction is also paralleled by a down-regulation of BDNF-activated signalling pathways in these cells (Carter *et al.*, 1995). Therefore, it is reasonable to assume that there exists an intracellular mechanism responsible for Trk receptor desensitization in the brain, and that such a mechanism could be present also in the visual cortex.

3.6   Neurotrophins and plasticity of the mammalian visual cortex: concluding remarks.

Results presented in this study clearly show that BDNF mRNA expression correlates to the onset of the period of plasticity in the rat visual cortex, and is dependent on visual experience both during and after this critical period. These findings open the question of which could be the role of BDNF (and other neurotrophins) in the plasticity of the rat visual cortex, also considering the large number of results published so far on this subject.

3.6.1 NGF, BDNF or other neurotrophic factors? An open question.

The large series of studies by the group of L. Maffei strongly support the hypothesis of a specific role of NGF in the development and plasticity of the rat visual cortex. However, some points remain unclear. First, neurons of dLGN do not express TrkA mRNA and it is not clear if they can retrogradely transport iodinated NGF when it is injected into the visual cortex (as the cholinergic afferents from basal forebrain do; Domenici et al., 1994b). According to this observation, dLGN neurons would not express the high affinity NGF receptor and geniculate fibres projecting to the visual cortex would not be able to transduce NGF signal. Therefore, it is unlikely that the effects of NGF on the visual cortex of monocularly deprived rats (see Introduction) are mediated by geniculo-cortical pathways. Some interpretations of these results are likely.

1) The action of NGF could be mediated by the cholinergic neurons of the basal forebrain, that are known to innervate the visual cortex (Carey and Rieck, 1987) and are supposed to exert a facilitatory action on the excitability of cortical neurons. However, a direct evidence for a role of cholinergic inputs in the plasticity of the rat visual cortex is still lacking.

2) NGF could have an intracortical action, since it has been reported that cortical neurons express TrkA mRNA.

3) Finally, also glial cells could be possible targets of NGF action, since they express both TrkA mRNA and protein (see above).

A second point concerns the regulation of NGF mRNA expression in the rat visual cortex. As I already mentioned, a major postulate of "neurotrophic hypothesis" is that neurotrophin factor should be produced by cortical neurons in an activity-dependent way. So far, there is no evidence that this is the case for NGF in the visual cortex. Castrén et al. (1992) failed to detect any change in NGF expression in dark-reared rats. In our experiments, we also could not detect any change in NGF expression in the rat visual cortex following monocular deprivation, but we ascribed this failure to a low sensitivity of in situ hybridisation protocol (Rossi, Bozzi, Pizzorusso, Maffei; unpublished observations). Conversely, BDNF mRNA expression is extremely sensitive to visual deprivation or stimulation, as widely shown by our results and by other Authors (see

Introduction), but a physiological role for BDNF in the plasticity of the rat visual system has not been demonstrated so far.

Finally, there is still the possibility that other neurotrophins could play a specific role in the segregation of geniculo-cortical connections in the mammalian visual cortex. This hypothesis is supported by the recent observation that exogenous administration of BDNF, NT-4/5 or antibodies against the BDNF and NT-4/5 receptor (TrkB) inhibit the formation of ocular dominance columns in the cat visual cortex (Cabelli *et al.*, 1995, and ref. 55 in Thoenen, 1995). Moreover, intracortical administration of NT-4/5 (but not NGF, BDNF or NT-3) prevents the shrinkage of dLGN neurons in monocularly deprived ferrets (Riddle *et al.*, 1995). The recent finding that NT-4/5 is mainly expressed in layer IV of the ferret (Cabelli *et al.*, 1995) and mouse (Bozzi, unpublished observation) visual cortex is consistent with a possible action of this neurotrophin on geniculo-cortical synapses. So far, as mentioned above for BDNF, nothing is known about the effects of NT-4/5 in the rat visual system.

It is evident from these results that different neurotrophins can exert different actions in the visual system of rats, cats and ferrets. These different actions might be certainly due to species differences. However, differences in the site and the methods of delivery of neurotrophic factors *in vivo* could be probably more important.

3.6.2   *A possible role for neurotrophins rises from recent studies on synaptic plasticity.*

A large series of recent findings strongly suggest a physiological role of neurotrophic factors in CNS neuronal plasticity. In particular, several studies over the past few years clearly demonstrate that neurotrophic factors have potentiating effects on excitatory synaptic transmission. An extensive review of these studies has been recently published by Thoenen (1995). The first evidence of such an action was given by Lohof *et al.* (1993), who demonstrated that both BDNF and NT-3 (but not NGF) rapidly potentiate the spontaneous and impulse-evoked synaptic activity of *Xenopus* neuromuscular junction in culture. More recently, Kang and Schuman (1995) have shown that application of BDNF and NT-3 causes a rapid and long-lasting (2-3 hours) enhancement of synaptic strength at the Schaffer collateral-CA1 synapses in hippocampal slices. This effect is blocked by the receptor tyrosine kinase inhibitor K252a, thus confirming that neurotrophin action is mediated by Trk receptors. Conversely, recent studies have demonstrated that hippocampal long-term potentiation is impaired in BDNF knockout mice (Korte *et al.*, 1995). LTP impairment is present not only in homozygous BDNF (-/-) mutants, but also in heterozygotes, in which the amount of BDNF mRNA in hippocampal neurons is reduced to ~50%. Therefore, BDNF gene dosage seems to be a limiting factor also for plasticity-related functions in the CNS. Consistent with these results, it has been recently demonstrated that BDNF can promote the induction of LTP by tetanic stimulation in hippocampal slices from young (postnatal day 12-13), which in the absence of BDNF can only undergo short-term potentiation (STP) of synaptic transmission (Figurov *et al.*, 1996). Finally, it has also been shown that BDNF can exert a potentiating effect on excitatory synaptic transmission in slices of visual cortex (Carmignoto *et al.*, 1993b).

Which could be the role for neurotrophic factors in CNS neuronal plasticity? Many recent findings suggest that they could act as activity-dependent "retrograde messengers" for synapse strengthening. A simple model based on a presynaptic neuron expressing Trk receptors and a postsynaptic target neuron producing a neurotrophin has been proposed (Thoenen, 1995; Bonhoffer, 1996; Ghosh, 1996; see Figure 3.1).

According to this model, afferent activity is required to induce neurotrophin expression and release from the postsynaptic neuron. The neurotrophin released in the synaptic cleft could then retrogradely act on presynaptic neuron to enhance neurotransmitter release and consequently strengthen the active synapse. For example, it haas been recently demonstrated that both BDNF and NGF can stimulate the phosphorylation of synapsin I in cerebrocortical neurons and PC12 cells (Jovanovic *et al.*, 1996), thus suggesting a direct role of neurotrophins in the regulation of neurotransmitter release by synapsin I. Moreover, many recent findings indeed demonstrate that NGF, BDNF and NT-4/5 can all enhance glutamatergic transmission in the hippocampus *in vitro* (see for instance Knipper *et al.*, 1994 and Leßmann *et al.*, 1994).

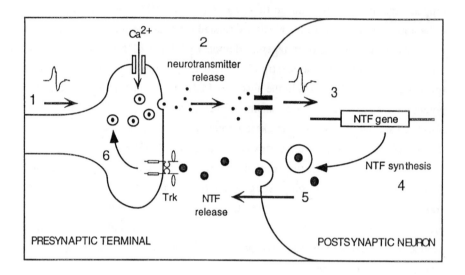

Figure 3.1.

A possible model for a role of neurotrophic factors as "retrograde messengers" in activity-dependent synaptic plasticity.

The model is based on a simple neuronal circuit, made by a presynaptic neuron expressing Trk receptors and a postsynaptic neuron producing one (or more) neurotrophic factor(s) in limiting amount. Action potential (1) induces Ca^{2+}-dependent neurotransmitter release (2) from the presynaptic terminal. The resulting excitation of the postsynaptic neuron, presumably through the activation of glutamate receptors, determines an increased expression (3), synthesis (4) and release (5) of neurotrophic factor(s) (NTF). Neurotrophin released in the synaptic cleft can retrogradely act on the presynaptic neuron via Trk receptors. This retrograde action could enhance the probability of neurotransmitter release (6) from the presynaptic terminal, thus stabilizing the active synapse.

The above model can likely apply to the process of synapse stabilization that takes place during the critical period in the mammalian visual cortex. Presynaptic dLGN (and cortical) neurons are known to express Trk neurotrophin receptors, whereas postsynaptic cortical neurons produce (limiting amount of) neurotrophic factors. The production and (presumably) release of neurotrophins in the visual cortex is finely regulated by afferent electrical activity. Neurotrophins secreted by postsynaptic cortical neurons could act as "retrograde messengers" on presynaptic geniculate (or cortical) neurons, thus enhancing neurotransmitter release from these cells. This would result in the stabilization of active synapses. So far, no data are available for such a role of neurotrophic factors in the visual system. In particular, nothing is known about the regulation of neurotransmitter release by neurotrophins in the visual cortex. Future experiments will contribute to elucidate this point.

TABLES OF FIGURES

Table I.

Developmental expression of BDNF mRNA in the rat visual cortex.

RNAase protection analysis of BDNF mRNA in rat visual cortex from postnatal (P) day 5 to adult (P60) age. The first two lanes on the left show native β-actin and BDNF probes. The bands indicated by asterisques represent the specific protection fragments for BDNF and β-actin mRNAs, respectively. MW: molecular weight markers (nucleotides, nt). Postnatal ages are indicated in days. **Methods**: Total RNAs were extracted from visual cortices from rats at different postnatal ages (P5-P60) as described (Bozzi et al., 1995), and RNAase protection assays were essentially performed following standard procedures (Sambrook et al., 1989). Thirty micrograms of total RNA per sample were hybridised with a molar excess of ^{32}P-labelled BDNF antisense RNA probe. The β-actin riboprobe was used in the same hybridisation mixture as an internal control for RNA quantification. After hybridisation, samples were treated with RNAase and analysed by electrophoresis followed by autoradiography. The protection bands shown here were obtained by different exposures of the same gel (4 and 24 hours for β-actin and BDNF, respectively).

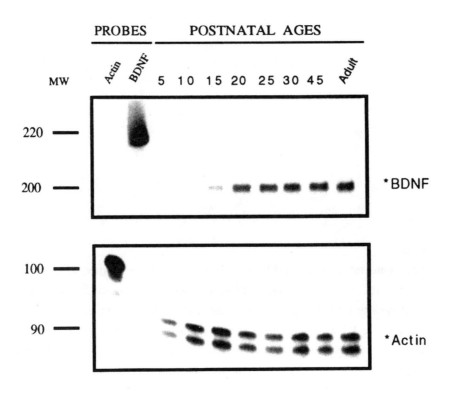

Table I.

Table II.

  Developmental increase of BDNF mRNA in the visual cortex, obtained by densitometric analysis of RNAase protection assays. Ratios of BDNF/β-actin optical density (O.D.) values (mean $\pm$ SEM of 3 different experiments) were normalized to P5 value and plotted as a function of the age of the analysed animals. The difference between P10 and P20 is statistically significant (Student's t-test, $p<0.05$). Open circles, postnatal ages between P5 and P45; closed circle, adults. **Methods:** The autoradiograms were analysed with a Molecular Dynamics 300A scanning densitometer. Optical density (O.D.) of BDNF and β-actin protection bands was calculated. β-actin was used as an internal standard. Due to the different amount of β-actin and BDNF mRNAs in the visual cortex, different exposures of the same gel were used (2, 4, 8, 16, 24, 32, 72 h). Quantification of protection bands was performed on films exposed for 4 h for β-actin and 24 h for BDNF, within the linear response range of the film for each band. Ratios of BDNF/β-actin O.D. values (mean $\pm$ SEM of 3 different experiments) were calculated and plotted.

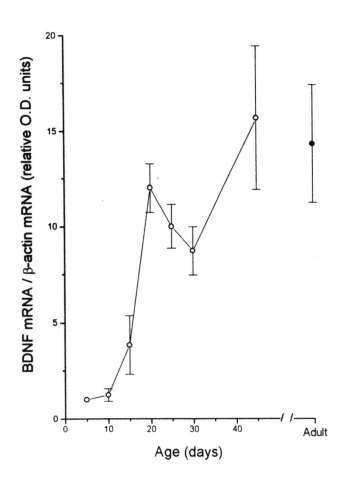

Table II.

Table III.

Developmental expression of BDNF mRNA in the rat visual cortex, determined by *in situ* hybridisation. (A, B, C) Dark-field images of coronal sections of the visual cortex, hybridised with a ^{35}S-labelled riboprobe for BDNF. Postnatal (P) ages are indicated in days. Scale bar = 600 μm. **Methods:** *In situ* hybridisation on cryostat sections was essentially performed as described (Bozzi *et al.*, 1995) following standard procedures (Wilkinson and Green, 1990).

P23

P45

P60

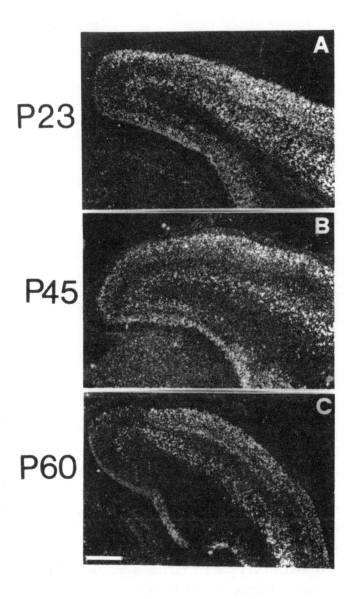

Table III.

Table IV.

Developmental expression of TrkB mRNA in the rat visual cortex.

RNAase protection analysis of TrkB mRNA in rat visual cortex from postnatal (P) day 5 to adult (P60) age. Total RNAs were hybridised with a ^{32}P-labelled TrkB riboprobe, and a ^{32}P-labelled β-actin riboprobe was used in the same hybridisation mixture as an internal control for RNA quantification. Native probes are not indicated. Exposure time: 24 hours. Postnatal ages are indicated in days. **Methods:** The rat TrkB cDNA was cloned by reverse transciption-polymerase chain reaction on total RNA extracted from brain of rats at postnatal day 20, using two oligos based on the sequence previously published (Middlemas *et al.*, 1991). The amplified 238 base pair (bp) fragment contained the tyrosine-kinase domain sequence (nucleotides 2163-2401 of rat TrkB sequence; TrkB-TK^{+}) and is specific for the full-length form of rat TrkB.

POSTNATAL AGE

5 10 15 20 30 45 A

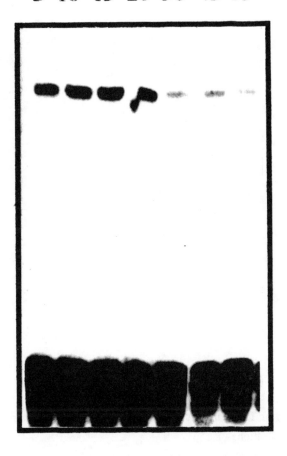

< trkB

< β-actin

Table IV.

YURI BOZZI

Table V.

Developmental expression of TrkB mRNA in the rat dLGN and visual cortex, determined by *in situ* hybridisation.

(A-D) Dark-field images of coronal sections of the visual cortex, hybridised with a ^{35}S-labelled TrkB-TK$^+$ riboprobe. (A'-D') The dark-field photomicrographs show coronal sections of dorsal lateral geniculate (dLGN) hybridised with the TrkB-TK$^+$ probe. The extension of dLGN in each section is delimited by white dashed lines. Postnatal ages are indicated in days. Scale bar = 600 μm.

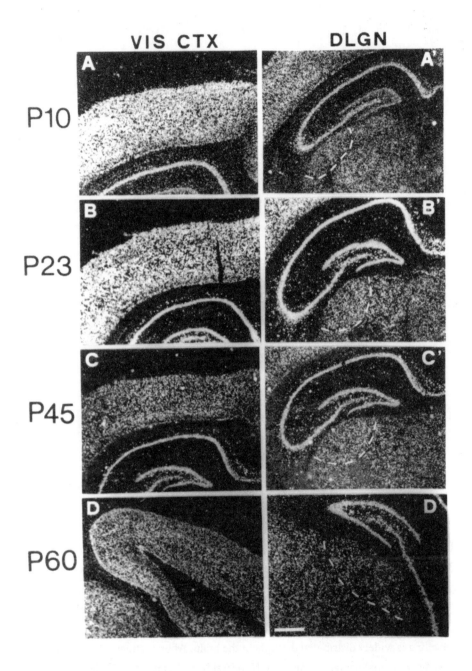

Table V.

Table VI.

BDNF and TrkB mRNA expression in the rat brain.

(A) Coronal section from a P45 rat brain hybridised with the BDNF ^{35}S-riboprobe. Neocortex, hippocampal dentate gyrus and mesencephalon are labelled. (B) Coronal section from a P30 rat brain hybridised with the TrkB-TK$^+$ ^{35}S-riboprobe. Labelling is widely distributed throughout the brain. Abbreviations: ctx, neocortex; DG, hippocampal dentate gyrus; dLGN, dorsal lateral geniculate (indicated by an arrowhead in B); hip, hippocampus; Pn, pontine nuclei. Scale bar = 1500 μm.

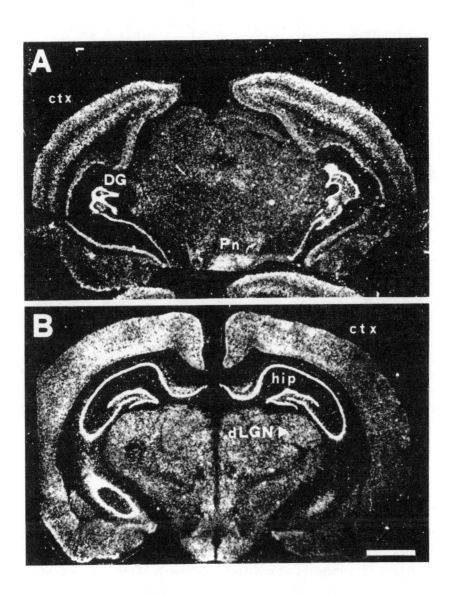

Table VI.

Table VII.

In situ hybridisation showing the localization of BDNF and TrkB mRNAs in the rat visual system at postnatal day 30, during the critical period.

(A, B) Dark-field images of a coronal section of the visual cortex, hybridised with ^{35}S-labelled riboprobes for BDNF and TrkB, respectively. Cortical layers are indicated by roman numbers. (C) The dark-field photomicrograph shows a coronal section of dorsal lateral geniculate (dLGN, indicated by arrow) hybridised with the TrkB probe. wm, white matter. Scale bar = 800 μm (A,C) and 700 μm (B).

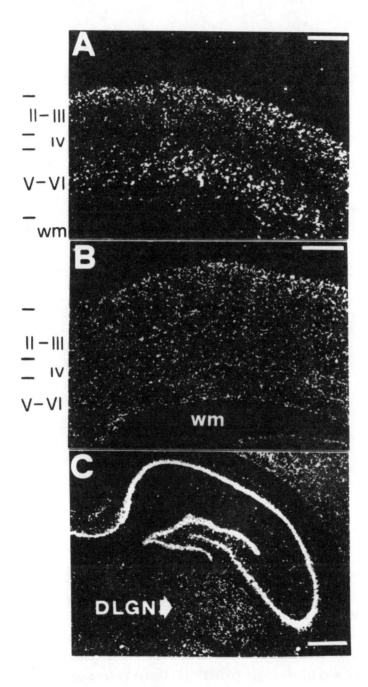

Table VII.

Table VIII.

BDNF and TrkB mRNA labelling in the layers of the visual cortex.

(A, B) Dark-field images of coronal sections of the visual cortex from a P30 brain, hybridised with BDNF (A) or TrkB (B). (C) Cresyl Violet counterstaining. Layers are indicated by numbers. wm, white matter. Scale bar = 400 μm.

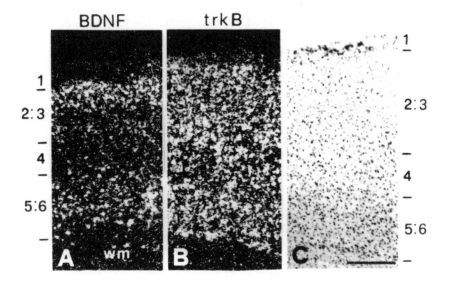

Table VIII.

Table IX.

BDNF and TrkB mRNA labelling on neurons of the visual cortex.

(A) Bright-field photomicrograph from a coronal section of P23 visual cortex (layer 5), hybridised with the BDNF probe. Some labelled neurons are indicated by arrows.

(B) In layers 2-3 of the visual cortex (P30), TrkB mRNA labellling is mainly restricted to neurons (indicated by an arrow). Some smaller (presumably glial) cells (indicated by a triangle) show a weaker labelling. Scale bar = 50 μm.

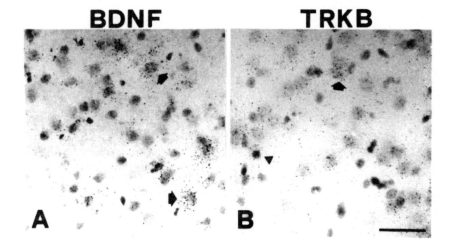

Table IX.

Table X.

BDNF immunoreactivity in the rat visual cortex at postnatal day P30.

(A) Coronal section of the visual cortex, probed with the anti-BDNF antibody. BDNF immunoreactivity (-IR) is mainly restricted to layers 2-3 and 5-6. Only few neurons in layer IV have a positive BDNF-IR. All positively stained cells are clearly neurons, as indicated by details from layers 2-3 (B) and 5-6 (C). BDNF-IR is confined only to the cytoplasm and dendrites, whereas nuclei are not labelled and appear white. Glial cells are not labelled. Scale bars: 300 μm (A) and 50 μm (B, C). **Methods:** Frozen cryostat sections (12-16 μm) were thawed at room temperature and rinsed for 2 h in 10% normal goat serum, 0.2% Triton X-100 in PBS to block non-specific binding. The primary rabbit anti-BDNF polyclonal antiserum was originally obtained by Dr Q. Yan (Amgen Corporation, USA) against the entire recombinant human BDNF. The BDNF antiserum was used at a concentration of 0,3 μg/ml in PBS, 10% normal goat serum, 0.2% Triton X-100, overnight at 4 °C. After a second blocking step, sections were incubated 1-2 h at room temperature with biotin-conjugated goat anti-rabbit IgG (Vector Laboratories, Burlingame, CA). Sections were then incubated for 1 h with avidin-biotin-peroxidase complex (Vectastain ABC kit, Vector Laboratories) and developed by diaminobenzidine/nichel method. The extension of primary visual cortex and monocular (Oc1M) and binocular (Oc1B) subfields was determined according to Paxinos and Watson (1986) and Zilles et al. (1984).

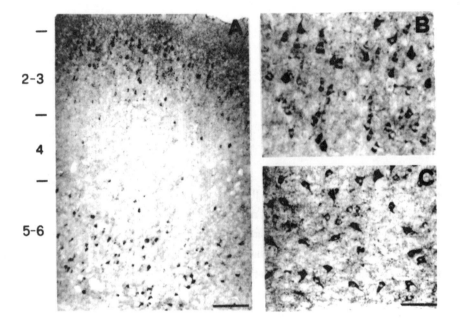

Table X.

Table XI.

Developmental expression of Trks in the rat visual cortex and basal forebrain.
Trk-like immunoreactivity (-IR) was analysed by immunoblot with an anti-panTrk
polyclonal antibody. Postnatal ages are indicated in days. Trk-like-IR in the visual
cortex is present before (P10) and during (P20-45) the critical period. In the basal
forebrain, Trks are already present at P10 and their expression slightly decreases
until P30. The molecular weight of Trk protein is indicated. **Methods:** Total
protein extracts were prepared from visual cortices and basal forebrains from rats at
different postnatal ages (P10-P45). Tissues were homogenised in 0.5-1 ml of lysis
buffer (150 mM NaCl, 20 mM Tris pH 8.0, 10% glycerol, 1% Triton X-100, 1mM
PMSF, 1 μg/ml aprotinin, 1 μg/ml leupeptin, and 4mM sodium orthovanadate) and
incubated for 20 min in ice. Lysates were centrifuged (13,000 r.p.m. for 30 min, at
4 °C) to eliminate cellular debris and protein concentration was estimated with the
Bradford method (Biorad). Thirty micrograms of total proteins/sample were
resolved by 7.5% SDS-PAGE. After blotting, filters were blocked and then probed
overnight at 4 °C with an anti-panTrk (C-14) rabbit polyclonal antibody (Santa
Cruz, CA, USA). Signal was then revealed by using a HRP-conjugated anti-rabbit
antibody (Biorad) and ECL (Amersham).

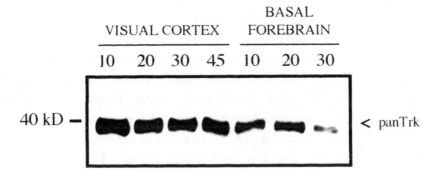

Table XI.

Table XII.

Monocular deprivation during the critical period reduces BDNF mRNA expression in the rat visual cortex.

(A,B) Dark field photomicrographs of *in situ* hybridisation with a BDNF ^{35}S-labelled riboprobe on coronal sections of the brain from a rat monocularly deprived from P15 to P45. (A) Visual cortex ipsilateral to the deprived eye. (B) Visual cortex contralateral to the deprived eye. The reduction of BDNF mRNA is restricted to the visual cortex contralateral to the deprived eye. Arrows in (B) delimitate the subfields of visual areas (Oc1B, binocular primary visual cortex. Oc1M, monocular primary visual cortex. Oc2L, lateral secondary visual cortex). (C) and (D) show magnifications of the visual cortex ipsilateral and contralateral to the deprived eye, respectively. The reduction of BDNF mRNA is more evident in superficial layers II/III. Cortical layers are indicated by roman numbers (wm, white matter). (E) *In situ* hybridisation on coronal section of the brain from a nondeprived P45 rat. No difference in BDNF mRNA expression is apparent between left and right visual cortex. Scale bar = 300 μm (A, B), 200 μm (C, D) and 1300 μm (E). **Methods:** Monocular deprivation was performed by means of eyelid suture, as described (Maffei *et al.*, 1992; Bozzi *et al.*, 1995). The effects of monocular deprivation on BDNF mRNAs were evaluated by comparing the labelling in the visual cortices ipsilateral and contralateral to the deprived eye. In the rat, left and right visual cortices receive afferents from both eyes, but the great majority (95-98%) of fibres from retinal ganglion cells crosses at the level of the optic chiasm (Sefton and Dreher, 1985). The classical organisation of the visual cortex in ocular dominance columns has not been observed in the rat (Zilles *et al.*, 1984). For these reasons, it is expected that the effects of the closure of one eye are predominantly restricted to the contralateral cortex. This criterion has been already followed by other Authors (Castrén *et al.*, 1992; Worley *et al.*, 1991) to evaluate changes in gene expression in the rat visual cortex induced by monocular injections of TTX. The extension and subdivision of visual cortical areas was evaluated by comparing sections subjected to *in situ* hybridisation with a series of adjacent sections counterstained with Cresyl Violet. Borders of primary visual cortex and monocular (Oc1M) and binocular (Oc1B) subfields were determined according to Paxinos and Watson (1986) and Zilles *et al.* (1984).

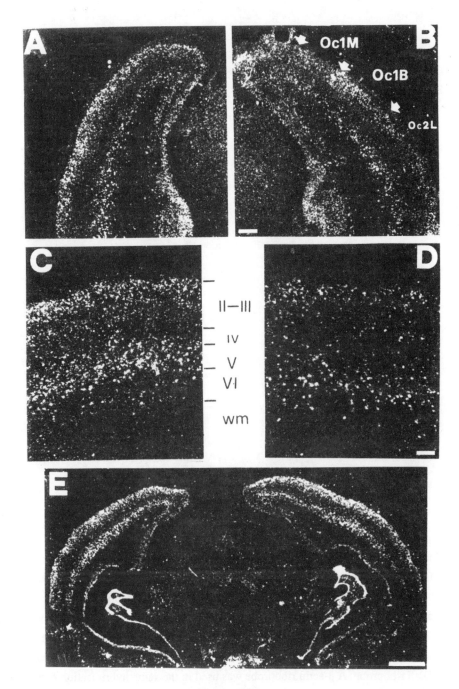

Table XII.

Table XIII.

Effect of monocular deprivation on BDNF mRNA expression in the visual cortex of rats during the critical period, determined by RNAase protection.

Rats were monocularly deprived from P15 to P30 as described (Bozzi *et al.*, 1995). Total RNAs were extracted from visual cortices ipsilateral (C, control) and contralateral (MD) to the deprived eye and analysed by RNAase protection with a BDNF riboprobe. A β-actin riboprobe was used in the same hybridisation mixture as an internal control.

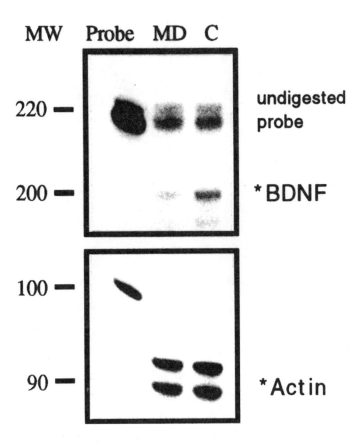

Table XIII.

Table XIV.

Effect of monocular deprivation on BDNF mRNA expression in the visual cortex of adult rats. (A, B, C) Dark field photomicrographs of *in situ* hybridisation with a BDNF ^{35}S-labelled riboprobe on coronal section of the brain from a rat monocularly deprived from P45 to P60. Arrows in (A, C) delimitate the extension of visual areas. The reduction of BDNF mRNA is restricted to the visual cortex contralateral to the deprived eye. (B) and (C) show magnifications of the visual cortex ipsilateral and contralateral to the deprived eye, respectively. The reduction of BDNF mRNA is more evident in superficial layers. DG, dentate gyrus. Oc1, primary visual cortex. Oc1B, binocular primary visual cortex. Oc1M, monocular primary visual cortex. Oc2L, lateral secondary visual cortex. Pn, pontine nuclei. Scale bar = 800 μm (A) and 600 μm (B,C).

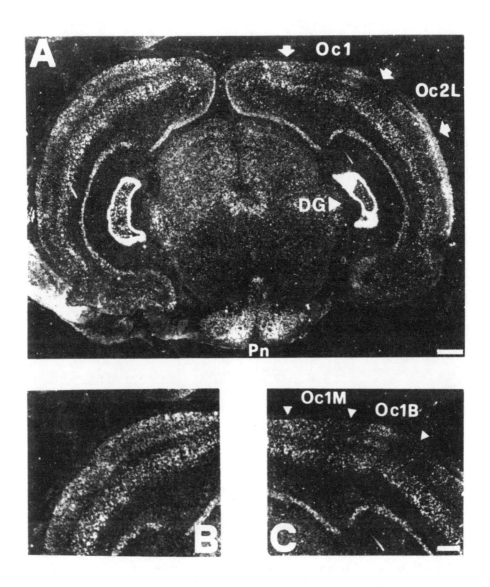

Table XIV.

Table XV.

Monocular deprivation does not affect TrkB mRNA expression in the rat dLGN and visual cortex. (A, B, C) Dark field photomicrographs of *in situ* hybridisation with a TrkB ^{35}S-labelled riboprobe on coronal sections of the brain from a rat monocularly deprived from P15 to P30. (A) TrkB mRNA is equally distributed through all cortical layers of the visual cortex ipsilateral (left) and contralateral (right) to the deprived eye. (B,C) TrkB mRNA labelling in the dLGN is not affected by MD. (B), ipsilateral side and (C) contralateral side to the deprived eye. Dorsal lateral geniculate nuclei (dLGN) are indicated by arrows. Scale bar = 960 μm (A) and 450 μm (B, C).

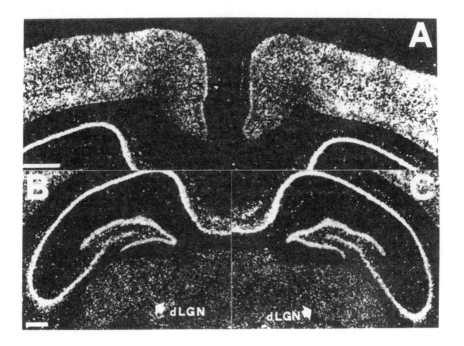

Table XV.

Table XVI.

Tetrodotoxin decreases the amount of BDNF mRNA in the visual cortex.

Rats received intraocular injections of TTX during or after the critical period as described (Bozzi *et al*, 1995). Total RNAs were extracted from visual cortices ipsilateral (C, control) and contralateral (TTX) to the injected eye and analysed by RNAase protection with a BDNF riboprobe. A β-actin riboprobe was used in the same hybridisation mixture as an internal control. Intraocular TTX injections reduced BDNF mRNA in the contralateral visual cortex with comparable efficacy in young (critical period, CP; 4 animals treated from from P15 to P30) and adult rats (8 animals treated from P45 to P50).

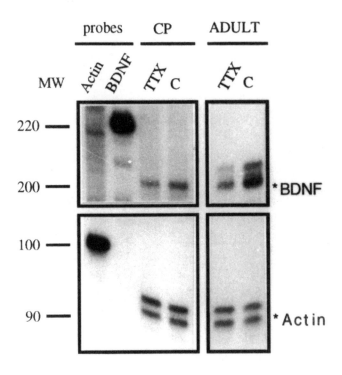

Table XVI.

Table XVII.

Bicuculline increases the amount of BDNF mRNA in the visual cortex.

Rats at postnatal day 30 received an intracortical administration of bicuculline for 6 hours as described (Bozzi *et al.*, 1995). (A) RNAase Protection. Lane BIC: RNA from bicuculline-treated visual cortex. Lane C: RNA from the contralateral control cortex. Lane N: RNA from the visual cortex of untreated rats of the same age.

(B) *In situ* hybridisation. The dark field photomicrograph shows a coronal section of visual cortex from a rat treated with bicuculline, hybridised with a ^{35}S-labelled riboprobe for BDNF. Slides were developed after 7 days of exposure to avoid saturation of BDNF labelling in the bicuculline-treated cortex. This procedure reduced the signal in the control cortex to background levels. Scale bar = 600 μm.

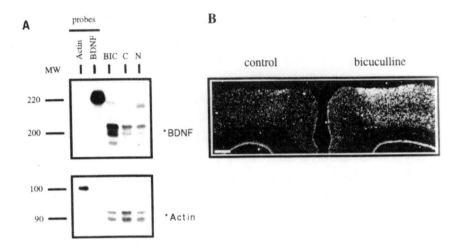

Table XVII.

Table XVIII. *In vitro* Trk phosphorylation in the visual cortex.

Visual cortices and basal forebrains from P20-21 rats were treated *in vitro* with NGF and BDNF as described below. (A) Lysates were immunoprecipitated with anti-panTrk antibody and immunoblotted with anti-phosphotyrosine (P-Tyr) antibody. In the basal forebrain (BF), no signal is present in control samples treated with Krebs' medium alone (CON, lane 2), whereas both NGF (lane 4) and BDNF (lane 6) induce Trk-tyrosine phosphorylation. In the visual cortex (VC), Trk-tyrosine phosphorylation is present both in untreated (CON, lane 1) and neurotrophin-treated (NGF, lane 3; BDNF, lane 5) samples. The band below Trks (approximately 90 kD) is probably an underglycosylated isoform of Trks and has been previously described also by other Authors (Li *et al.*, 1995). (B) To quantify Trk-like proteins, the same filter as in (A) was stripped and reprobed with the anti-panTrk antibody. All samples showed a comparable signal at 140 kD, corresponding to Trk. MW: molecular weight marker. **Methods:** Six 20-21 days old-rats were used for *in vitro* neurotrophin treatment of freshly dissected tissues. Two animals were used for NGF treatment, two for BDNF treatment and two for control experiments. Rats were anaesthetized with chloral hydrate, and brains were quickly removed and dissected. First, the occipital part of the neocortex (visual cortex) was collected. Next, basal forebrain was dissected. After dissection, tissues were immediately cut in smaller pieces ("microprisms": thickness, ~250 μm; weight, 10-100 mg) and transferred in Eppendorf tubes containing 1.5 ml of 5% CO_2-equilibrated Krebs' medium and processed as described (Knüsel *et al.*, 1994; Li *et al.*, 1995). NGF (kindly provided by Dr. D.Mercanti, CNR Institute of Neurobiology, Rome, Italy) and BDNF (a gift of Dr. E. Castrén, MPI for Psychiatry, Martinsried, Germany) were added at a final concentrations of 100 and 300 ng/ml, respectively, and samples were incubated 5 min at 37 °C. As control, tissues from two animals were incubated in Krebs' solution with no addition of neurotrophins. After treatment, tubes were immediately placed on ice, quickly centrifuged and the supernatant removed. Pellets were were homogenized in 1 ml of lysis buffer (see legend of Table XI) and incubated for 20 min in ice. Lysates were centrifuged (13,000 r.p.m. for 30 min, at 4 °C) to eliminate cellular debris, transferred to fresh tubes and stored at -80 °C until use for immunoprecipitation and western blot with anti-panTrk (Santacruz Biotech., CA, USA) and anti-phosphotyrosine (Upstate Biotech., NY, USA) antibodies.

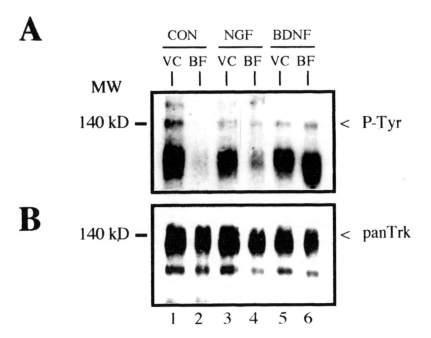

Table XVIII.

Table XIX. *In vivo* Trk phosphorylation in the visual cortex.

NGF and BDNF were locally applied for 30 min onto the right visual cortex of 23-24 day-old rats. Left visual cortex was treated with saline solution as control. Tissues were homogenized and immunoprecipitated with an anti-panTrk antibody and immunoblotted with an anti-P-Tyr antibody. (A) Neither NGF nor BDNF can increase the basal level of Trk-tyrosine phosphorylation in the visual cortex. A 140 kD band corresponding to phosphorylated Trks is both present in NGF (lane 1) or saline (lane 2) treated visual cortices. A comparable level of Trk-tyrosine phosphorylation is present in untreated frontal cortices dissected from the same animal (lanes 3 and 4). No increase in Trk-tyrosine phosphorylation was present in the BDNF-treated VC (lane 5), as compared with saline-treated visual cortex (lane 6) or untreated frontal cortices (lane 7). The weaker signal in lane 8 is probably due to sample degradation, as also demonstrated by the weaker Trk band in B, lane 8. (B) Reprobing the same filter as in (A) with the anti-panTrk antibody demonstrated that the level of Trks was comparable in all samples (except in lane 8, that contains a degraded sample). MW: molecular weight markers. The strong band at 50 kD indicated by IgG in (A) and (B) corresponds to the heavy chain of the anti-panTrk antibody used for immunoprecipitation, recognized by the anti-rabbit immunoglobulin (IgG) secondary antibody. **Methods:** Neurotrophin treatment was performed under chloral hydrate anaesthesia on 4 rats at P23-24. Two rats were treated with NGF and two with BDNF. Neurotrophins were applied directly onto the right visual cortex after surgical removal of the dural membrane. Local application was performed with a small piece (~1 mm^2) of fibrine (Zimospuma, 80% fibrine, 20% NaCl, Baldacci Laboratories, Pisa, Italy). Fibrine was soaked with 3-5 µl of a 1.5 mg/ml solution of either NGR or BDNF. As control, vehicle solution was applied onto the left cortex. Rats were killed 30 min after the treatment. First, visual cortices were rapidly dissected. Then, frontal (untreated) cortices were then removed. Tissues were frozen and stored at -80 °C until use for Trk tyrosine phosphorylation analysis.

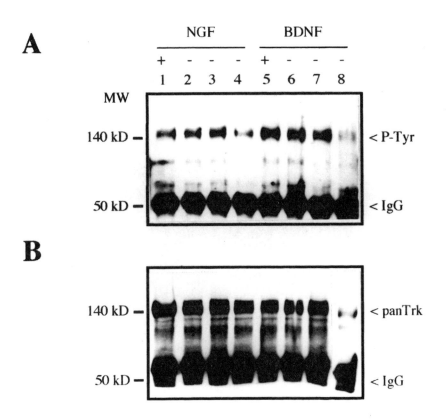

Table XIX.

REFERENCES

Acheson A., Barker P.A., Alderson R.F., Miller F.D. and Murphy R.A. (1991) Detection of Brain-derived Neurotrophic Factor-like activity in fibroblasts and Schwann cells: inhibition by antibodies to NGF. *Neuron* **7**, 265-275.

Alderson R.F., Alterman A.L., Barde Y.A. and Lindsay R.M. (1990) Brain-derived neurotrophic factor increases survival and differentiated functions of rat septal cholinergic neurons in culture. *Neuron* **5**, 297-306.

Allendoerfer K.L., Cabelli R.J., Escandón E., Kaplan D.R., Nikolics K. and Shatz C.J. (1994) Regulation of neurotrophin receptors during the maturation of the mammalian visual system. *J. Neurosci.* **14**, 1795-1811.

Angeletti R.H. and Bradshaw, R.A. (1971) Nerve growth factor from mouse submaxillary gland: amino acid sequence. *Proc. Natl. Acad. Sci. USA* **65**, 2417-2420.

Baldwin A.N., Bitler C.M., Welcher A.A. and Shooter E.M. (1992) Studies on the structure and binding properties of the cysteine-rich domain of rat low affinity nerve growth factor receptor (p75NGFR). *J. Biol. Chem.* **267**, 8352-8359.

Bandtlow C., Heumann R., Schwab M.E. and Thoenen H. (1987) Cellular localization of Nerve Growth Factor synthesis by *in situ* hybridization. *EMBO J.* **6**, 891-899.

Barbacid M. (1995) Neurotrophic factors and their receptors. *Curr. Opin. Cell Biol.* **7**, 148-155.

Barde Y.A. (1989) Trophic factors and neuronal survival. *Neuron* **2**, 1525-1534.

Barde Y.A., Edgar D. and Thoenen H. (1982) Purification of a new neurotrophic factor from mammalian brain. *EMBO J.* **1**, 549-553.

Barker P.A. and Shooter E.M. (1994) Disruption of NGF binding to the low affinity neurotrophin receptor p75^{LNTR} reduces NGF binding to TrkA on PC12 cells. *Neuron* **13**, 203-215.

Bear M.F., Kleinschmidt A., Gu Q. and Singer W. (1990) Disruption of experience-dependent synaptic modifications in striate cortex by infusion of an NMDA receptor antagonist. *J. Neurosci.* **10**, 909-925.

Berardi N., Cellerino A., Domenici L., Fagiolini M., Pizzorusso T., Cattaneo A. and Maffei L. (1994) Monoclonal antibodies to nerve growth factor affect the postnatal development of the visual system. *Proc. Natl. Acad. Sci. USA* **91**, 684-688.

Berkemeier L., Winslow J., Kaplan D., Nicolics K., Goeddel D. and Rosenthal A. (1991) Neurotrophin-5: a novel neurotrophic factor that activates trk and trkB. *Neuron* **7**, 857-866.

Bernd P. and Greene L.A. (1984) Association of ^{125}I-nerve growth factor with PC12 pheochromocytoma cells. *J. Biol. Chem.* **259**, 15509-15516.

Berninger B., Marty S., Zafra F., Berzaghi M.P., Thoenen H. and Lindholm D. (1995) GABAergic stimulation switches from enhancing to repressing BDNF expression in rat hippocampal neurons during maturation in vitro. *Development* **121**, 2327-2335.

Berzaghi M.P., Cooper J., Castrén E., Zafra F., Sofroniew M., Thoenen H. and Lindholm D. (1993) Cholinergic regulation of brain-derived neurotrophic factor (BDNF) and nerve growth factor (NGF) but not neurotrophin-3 (NT-3) mRNA levels in the developing rat hippocampus. *J.Neurosci.* **13**, 3818-3826.

Berzaghi M.P., Freund T., Papp E., Castrén E., Zirrgiebel U., Thoenen H. and Lindholm D. (1994) Co-localization of Brain-derived Neurotrophic Factor (BDNF) and parvalbumin in the rat cerebral cortex. *Soc. Neurosci. Abstr.* **20**, 536.14.

Bliss T.V.P. and Collingridge G.L. (1993) A synaptic model of memory: long-term potentiation in the hippocampus. *Nature* **361**, 31-39.

Bliss T.V.P. and Lomo (1973) Long-term potentiation of synaptic transmission in the dentate area of the anaesthetized rabbit following stimulation of the perforant path. *J. Physiol (Lond.)* **232**, 331-356.

Blöchl A. and Thoenen H. (1995) Characterization of nerve growth factor (NGF) release from hippocampal neurons: evidence for a constitutive and an unconventional sodium-dependent regulated pathway. *Eur. J. Neurosci.* **7**, 1220-1228.

Bonhoffer T. (1996) Neurotrophins and activity-dependent development of the neocortex. *Curr. Op. Neurobiol.* **6**, 119-126.

Bonni A., Ginty D.D., Dudek H. and Greenberg M.E. (1995) Serine 133-phosphorylated CREB induces transcription via a cooperative mechanism that may confer specificity to neurotrophin signals. *Mol. Cell. Neurosci.* **6**, 168-183.

Bothwell M. (1995) Functional interactions of neurotrophins and neurotrophin receptors. *Annu. Rev. Neurosci.* **18**, 221-253.

Bozzi Y., Pizzorusso T., Cremisi F., Rossi F.M., Barsacchi G. and Maffei L. (1995) Monocular deprivation decreases the expression of messenger RNA for brain-derived neurotrophic factor in the rat visual cortex. *Neuroscience* **69**, 1133-1144.

Cabelli R.J., Hohn A. and Shatz C.J. (1995a) Inhibition of ocular dominance formation by infusion of NT-4/5 or BDNF. *Science* **267**, 1662-1666.

Cabelli R.J., Tavazoie S. and Shatz C.J. (1995b) Changing patterns of BDNF and NT-4/5 immunoreactivity during visual system development. *Soc. Neurosci Abstr.* **21**, 706.16.

Carey R.G. and Rieck R.W. (1987) Topographic projections to the visual cortex from the basal forebrain of the rat. *Brain Res.* **424**, 205-215.

Carmignoto G., Comelli M.C., Candeo P., Cavicchioli L., Yan Q., Merighi A. and Maffei L. (1991) Expression of NGF receptor and NGF receptor mRNA in the developing and adult rat retina. *Exp. Neurol.* **111**, 302-311.

Carmignoto G. and Vicini S. (1992) Activity-dependent decrease in NMDA receptor responses during development of the visual cortex. *Science* **258**, 1007-1011.

Carmignoto G., Canella R., Candeo P., Comelli M.C. and Maffei L. (1993a) Effects of NGF on neuronal plasticity of the kitten visual cortex. *J. Physiol. (London)* **464**, 343-360.

Carmignoto G., Negro A. and Vicini S. (1993b) NGF and BDNF modulate excitatory synapses in rat visual cortical neurons. *Soc. Neurosci. Abstr.* **19**, 690.9.

Carter B.D., Zirrgiebel U. and Barde Y.-A. (1995) Differential regulation of p21^{ras} activation in neurons by Nerve Growth Factor and Brain-derived Neurotrophic Factor. *J. Biol. Chem.* **270**, 21751-21757.

Carter B.D., Kaltschmidt C., Kaltschmidt B., Offenhauser N., Bohm-Matthaei R., Bauerle P.A. and Barde Y.-A. (1996) Selective activation of NF-kB by nerve growth factor through the neurotrophin receptor p75. *Science* **227**, 542-545.

Castrén E., Zafra F., Thoenen H. and Lindholm D. (1992) Light regulates expression of brain-derived neurotrophic factor mRNA in the rat visual cortex. *Proc. Natl. Acad. Sci. USA* **89**, 9444-9448.

Castrén E., Berninger B., Leingärtner A., Lindholm D. and Thoenen H. (1995) Transcriptional regulation of brain-derived neurotrophic factor mRNA levels in hippocampus by kainic acid is independent of protein synthesis and activation of calmoduline kinase IIa. *Soc. Neurosci. Abstr.* **21**, 417.7.

Cellerino A., Burkhalter A., Maffei L. and Domenici L. (1995) Expression of TrkB but not BDNF mRNA in parvalbumin containing neurons of the adult rat visual cortex. *Soc. Neurosci. Abstr.* **21**, 790.7.

Cellerino A. and Maffei L. (1996) The action of neurotrophins in the development and plasticity of the visual cortex. *Prog. Neurobiol.* **49**, 53-71.

Chandler C.E., Parsons L.M., Hosang M. and Shooter E.M. (1984) A monoclonal antibody modulates the interaction of nerve growth factor with PC12 cells. *J. Biol. Chem.* **259**, 6882-6889.

Chao M.V., Bothwell M.A., Ross A.H., Koprowski H., Lanahan A.A., Buck C.R. and Sehgal A. (1986) Gene transfer and molecular cloning of human NGF receptor. *Science* **232**, 518-521.

Chao M.V. and Hempstead B.L. (1995) p75 and Trk: a two-receptor system. *Trends Neurosci.* **18**, 321-326.

Chaudhuri A. and Cynader M.S. (1993) Activity-dependent expression of the transcription factor Zif268 reveals ocular dominance columns in monkey visual cortex. *Brain Res.* **605**, 349-353.

Choen S. (1960) Purification of a nerve-promoting protein from the mouse salivary gland and its neurocytotoxic antiserum. *Proc. Natl. Acad. Sci. USA* **46**, 302-311.

Clary D.O. and Reichardt L.F. (1994) An alternatively spliced form of the nerve growth factor receptor TrkA confers an enhanced response to neurotrophin-3. *Proc. Natl. Acad. Sci. USA* **91**, 11133-11137.

Clary D.O., Weskamp G., Austin L.R. and Reichardt L.F. (1994) TrkA cross-linking mimics neuronal responses to nerve growth factor. *Mol. Biol. Cell* **5**, 549-563.

Conover J.C., Erickson J.T., Katz D.M., Bianchi L.M., Pouemyrou W.T., McClain J., Pan L., Heigren M., Ip N.Y., Boland P., Friedman B., Wiegand S., Vejsada R., Kato A.C., DeChiara T.M. and Yancopoulos G.D. (1995) Neuronal deficits, not involving motor neurons, in mice lacking BDNF and/or NT4. *Nature* **375**, 235-238.

Cramer K.S. and Sur M. (1995) Activity-dependent remodeling of connections in the mammalian visual system. *Curr. Opin. Neurobiol.* **5**, 106-111.

Crowley C., Spencer S.D., Nishimura N.C., Chen K.S., Pitts-Meek S., Armanini M.P., Ling L.H., McMahon S.B., Shelton D.L., Levinson A.D. and Phillips H.D. (1994) Mice lacking Nerve Growth Factor display perinatal loss of sensory and sympathetic neurons yet develop basal forebrain cholinergic neurons. *Cell* **76**, 1001-1011.

Curtis R., Adryan K.M., Stark J.L., Park J.S., Compton D.L., Weskamp G., Huber L.J., Chao M.V., Jaenisch R., Lee K.-F., Lindsay R.M. and DiStefano P.S. (1995) Differential role of low affinity neurotrophin receptor (p75) in retrograde axonal transport of the neurotrophins. *Neuron* **14**, 1201-1211.

Davis R.J. (1993) The mitogen-activated protein kinase signal transduction pathway. *J. Biol. Chem.* **268**, 14553-14556.

DiStefano P.S., Friedmann B., Radziejewski C., Alexander C., Boland P., Schick C.M., Lindsay R.M. and Wiegand S.J. (1992) The neurotrophins BDNF, NT-3, and NGF display distinct pattern of retrograde axonal transport in peripheral and central neurons. *Neuron* **8**, 983-993.

Dobrowsky R.T., Werner M.H., Castellino A.M., Chao M.V. and Hannun Y.A. (1994) Activation of the sphingomyelin cycle through the low-affinity neurotrophin receptor. *Science* **265**, 1596-1599.

Domenici L., Berardi N., Carmignoto G., Vantini G. and Maffei L. (1991) Nerve growth factor prevents the amblyopic effects of monocular deprivation. *Proc. Natl. Acad. Sci. USA* **88**, 8811-8815.

Domenici L., Cellerino A., and Maffei L. (1993) Monocular deprivation effects in the rat visual cortex and lateral geniculate nucleus are prevented by nerve growth factor (NGF). II: Lateral geniculate nucleus. *Proc. R. Soc. Lond. B* **251**, 25-31.

Domenici L., Cellerino A., Berardi N., Cattaneo A. and Maffei L. (1994a) Antibodies to nerve growth factor (NGF) prolong the sensitive period for monocular deprivation in the rat. *NeuroReport* **5**, 2041-2044.

Domenici L., Fontanesi G., Cattaneo A., Bagnoli P. and Maffei L. (1994b) Nerve Growth Factor uptake and transport folowing injection in the developing rat visual cortex. *Vis. Neurosci.* **11**, 1093-1102.

Dugich-Djordjevic M.M., Tocco G., Willoughby D.A., Najm I., Pasinetti G., Thompson R.F., Baudry M., Lapchak P.A. and Hefti F. (1992) BDNF mRNA expression in the developing rat brain following kainic acid-induced seizure activity. *Neuron* **8**, 1127-1138.

Dugich-Djordjevic M.M., Peterson C., Isono F., Ohsawa F., Widmer H.R., Denton T.L., Bennet G. and Hefti F. (1995) Immunohistochemical localization of Brain-derived Neurotrophic Factor in the brain. *Eur. J. Neurosci.* **7**, 1831-1839.

Ernfors P., Wetmore C., Olson L. and Persson H. (1990) Identification of cells in rat brain and peripheral tissues expressing mRNA for members of the nerve growth factor family. *Neuron* **5**, 511-526.

Ernfors P., Bengzon J., Kokaia Z., Persson H. and Lindvall O. (1991) Increased levels of messenger RNAs for neurotrophic factors in the brain during kindling epileptogenesis. *Neuron* **7**, 165-176.

Ernfors P., Lee K.-F. and Jaenisch R. (1994a) Mice lacking brain-derived neurotrophic factor develop with sensory deficits. *Nature* **368**, 147-150.

Ernfors P., Lee K.-F., Kucera J. and Jaenisch R. (1994b) Lack of neurotrophin-3 leads to deficiencies in the peripheral nervous system and loss of limb proprioceptive afferents. *Cell* **77**, 503-512.

Fagiolini M., Pizzorusso T., Berardi N., Domenici L. and Maffei L. (1994a) Functional postnatal development of the rat primary visual cortex and the role of visual experience: dark rearing and monocular deprivation. *Vision Res.* **34**, 709-720.

Fagiolini M., Pizzorusso T., Cenni M.C., Porciatti V. and Maffei L. (1994b) Visual deficits induced by dark rearing are prevented by Schwann cell transplant in the rat. *Soc. Neurosci. Abstr.* **20**, 203.3.

Fariñas I., Jones K.R., Backus C., Wang X.Y. and Reichardt L.F. (1994) Severe sensory and sympathetic deficits in mice lacking neurotrophin-3. *Nature* **369**, 658-661.

Figurov A., Pozzo-Miller L.D., Olafsson P., Wang T. and Lu B. (1996) Regulation of synaptic responses to high-frequency stimulation and LTP by neurotrophins in the hippocampus. *Nature* **381**, 706-709.

Fiorentini A., Berardi N. and Maffei L. (1995) Nerve Growth Factor preserves behavioral visual acuity in monocularly deprived kittens. *Vis. Neurosci.* **12**, 51-55.

Fox K. and Zahs K. (1994) Critical period control in sensory cortex. *Curr. Opin. Neurobiol.* **4**, 112-119.

Fox K. (1995) The critical period for long-term potentiation in primary sensory cortex. *Neuron* **15**, 485-488.

Fraizer W.A., Boyd L.F. and Bradshaw R.A. (1974) Properties of the specific binding of ^{125}I-nerve growth factor to responsive peripheral neurons. *J. Biol. Chem.* **249**, 5513-5519.

Frisén J., Verge V.M.K., Fried K., Risling M., Persson H., Trotter J., Hökfelt T. and Lindholm D. (1995) Characterization of glial trkB receptors: differential response to injury in the central and peripheral nervous systems. *Proc. Natl. Acad. Sci. USA* **90**, 4971-4975.

Gall C.M. and Isackson P.J. (1989) Limbic seizures increase neuronal production of mRNA for nerve growth factor. *Science* **245**, 758-761.

Gall C.M., Murray K. and Isackson P.J. (1991) Kainic acid-induced seizures increased expression of nerve growth factor mRNA in hippocampus. *Mol. Brain Res.* **9**, 113-123.

Ghosh A., Carnahan J. and Greenberg M.E. (1994) Requirement for BDNF in activity-dependent survival of cortical neurons. *Science* **263**, 1618-1623.

Ghosh A. (1996) With an eye on neurotrophins. *Curr. Biol.* **6**, 130-133.

Gibbs R.B. and Pfaff D.W. (1994) In situ hybridization detection of trkA mRNA in brain: distribution, colocalization with p75^{NGFR} and up-regulation by nerve growth factor. *J. Comp. Neur.* **341**, 324-339.

Ginty D.D., Kornhauser J.M., Thompson M.A., Bading H., Mayo K.E., Takahashi J.S. and Greenberg M.E. (1993) Regulation of CREB phosphorylation in the suprachiasmatic nucleus by light and a circadian clock. *Science* **260**, 238-241.

Ginty D.D., Bonni A. and Greenberg M.E. (1994) Nerve growth factor activates a ras-dependent protein kinase that stimulates *c-fos* transcription via phosphorylation of CREB. *Cell* **77**, 713-725.

Glass D.J. and Yancopoulos G.D. (1993) The neurotrophins and their receptors. *Trends Cell Biol.* **3**, 262-268.

Godfrey E.W. and Shooter E.M. (1986) Nerve growth factor receptors on chick embryo sympathetic ganglion cells: binding characteristics and development. *J. Neurosci.* **6**, 2543-2550.

Götz R., Köster R., Winkler C., Raulf F., Lottspeich F., Schartl M. and Thoenen H. (1994) Neurotrophin-6 is a new member of the nerve-growth factor family. *Nature* **372**, 266-269.

Greene L.A. and Kaplan D.R. (1995) Early events in neurotrophin signalling via Trk and p75 receptors. *Curr. Opin. Neurobiol.* **5**, 579-587.

Greene L.A. and Tischler A.S. (1976) Establishment of a noradrenergic clonal line of rat adrenal pheochromocytoma cells which respond to nerve growth factor. *Proc. Natl. Acad. Sci. USA* **73**, 2424-2428.

Griesbeck O., Blöchl A., Carnahan J.F., Nawa H. and Thoenen H. (1995) Characterization of Brain-derived neurotrophic (BDNF) secretion from hippocampal neurons. *Soc. Neurosci. Abstr.* **21**, 417.12.

Grob P.M., Berlot C.H. and Bothwell M.A. (1983) Affinity labeling and partial purification of nerve growth factors receptors from rat pheochromocytoma and human melanoma cells. *Proc. Natl. Acad. Sci. USA* **80**, 6819-6823.

Hallböök F., Ibañez C.F. and Persson H. (1991) Evolutionary studies of the nerve growth factor family reveal a novel member abundantly expressed in Xenopus ovary. *Neuron* **6**, 845-858.

Hayashi M., Yamashita A. and Shimizu K. (1990) Nerve Growth Factor in the primate central nervous system: regional distribution and ontogeny. *Neuroscience* **36**, 683-689.

Hebb D.O. (1949) *The organization of behavior.* John Wiley and Sons, Inc., New York.

Hefti F. (1986) Nerve growth factor (NGF) promotes survival of septal cholinergic neurons after fimbria transections. *J.Neurosci.* **6**, 2155-2162.

Hendry S.H.C. and Jones E.G. (1986) Reduction in number of immunostained GABAergic neurones in deprived-eye dominance columns of monkey area 17.

Nature **320**, 750-753.

Hendry S.H.C. and Kennedy M.B. (1986) Immunoreactivity for a calmodulin-dependent protein kinase is selectively increased in macaque striate cortex after monocular deprivation. *Proc. Natl. Acad. Sci. USA* **83**, 1536-1540.

Hengerer B., Lindholm D., Heumann R., Ruther U., Wagner E.F. and Thoenen H. (1990) Lesion-induced increase in nerve growth factor mRNA is mediated by c-fos. *Proc. Natl. Acad. Sci. USA* **87**, 3899-3903.

Heumann R. (1994) Neurotrophin signalling. *Curr. Opin. Neurobiol.* **4**, 668-679.

Hofer M., Pagliusi S.R., Hohn A., Leibrock J. and Barde Y.-A. (1990) Regional distribution of brain-derived neurotrophic factor mRNA in the adult mouse brain. *EMBO J.* **9**, 2459-2464.

Hoffman D., Breakefield X.O., Short M.P. and Aebischer P. (1993) Transplantation of polymer-encapsulated cell line genetically engineered to release NGF. *Exp. Neurol.* **122**, 100-106.

Hohn A., Leibrock J., Bailey K. and Barde Y.A. (1990) Identification and characterization of a novel member of the nerve growth factor/brain-derived neurotrophic factor family. *Nature* **344**, 339-341.

Holtzman D.M., Li Y., Parada L.F., Kinsman S., Chen C.-K., Valletta J.S., Zhou J., Long J.B., Mobley W.C. (1992) p140trk mRNA marks NGF-responsive forebrain neurons: evidence that trk gene expression is induced by NGF. *Neuron* **9**, 465-478.

Hosang M. and Shooter E.M. (1985) Molecular characteristics of nerve growth factor receptors on PC12 cells. *J. Biol. Chem.* **260**, 655-662.

Hubel D.H. and Wiesel T.N. (1963) Shape and arrangement of columns in cat's striate cortex. *J. Physiol.* **165**, 559-568.

Hubel D.H. and Wiesel T.N. (1965) Binocular interaction in striate cortex of kittens reared with artificial squint. *J. Neurophysiol.* **28**, 1041-1059.

Hubel D.H. and Wiesel T.N. (1970) The period of susceptibility to the physiological effects of unilateral eye closure in kittens. *J. Physiol.* **206**, 419-436.

Hubel D.H., Wiesel T.N. and LeVay S. (1977) Plasticity of ocular dominance columns in the monkey striate cortex. *Phil. Trans. R. Soc. (London)* **B 278**, 377-409.

Hyman C., Hofer M., Barde Y.-A., Jurhasz M., Yancopoulos G., Squinto S.P. and Lindsay R.M. (1991) BDNF is a neurotrophic factor for dopaminergic neurons of the substantia nigra. *Nature* **350,** 230-232.

Ip N., Ibañez C.F., Nye S., Mc Clain J., Jones P., Gies D., Belluscio L., Le Beau M.M., Espinosa III M., Squinto S., Persson H. and Yancopoulos G. (1992) Mammalian neurotrophin-4: structure, distribution and receptor specificity. *Proc. Natl. Acad. Sci. USA* **89**, 3060-3064.

Isackson P.J. (1995) Trophic factor response to neuronal stimuli or injury. *Curr. Opin. Neurobiol.* **5**, 350-357.

Isackson P.J., Huntsmann M.M., Murray K.D. and Gall C.M. (1991) BDNF mRNA expression is increased in adult rat forebrain after limbic seizures: temporal patterns of induction distinct from NGF. *Neuron* **6**, 937-948.

Jing S, Tapley P. and Barbacid M. (1992) Nerve growth factor mediates signal transduction through trk homodimer receptors. *Neuron* **9**, 1067-1079.

Jones K.R., Fariñas I., Backus C. and Reichardt L.F. (1994) Targeted disruption of BDNF gene perturbs brain and sensory neuron development but not motor neuron development. *Cell* **76**, 989-999.

Jovanovic J.N., Benfenati F., Siow Y.L., Sihra T.S., Sanghera J.S., Pelech

S.L., Greengard P. and Czernik A.J. (1996) Neurotrophins stimulate phosphorylation of synapsin I by MAP kinase and regulate synapsin I-actin interactions. *Proc. Natl. Acad. Sci.USA* **93**, 3679-3683.

Kageyama G.H. and Robertson R.T. (1993) Development of geniculocortical projections to visual cortex in rat: evidence for early ingrowth and synaptogenesis. *J. Comp. Neurol.* **335**, 123-148.

Kang H. and Schuman E.M (1995) Long-lasting neurotrophin induced enhancement of synaptic transmission in the adult hippocampus. *Science* **267**, 1658-1662.

Kaplan D.R., Hempstead B.L., Martin-Zanca D., Chao M.V. and Parada L.F. (1991a) The trk proto-oncogene product: a signaling transducing receptor for nerve growth factor. *Science* **252**, 554-557.

Kaplan D.R., Martin-Zanca D. and Parada L.F. (1991b) Tyrosine phosphorylation and tyrosine kinase activity of the trk proto-oncogene product induced by NGF. *Nature* **350**, 158-160

Kirkwood A., Lee H.K. and Bear M.F. (1995) Co-regulation of long-term potentiation and experience-dependent synaptic plasticity in visual cortex by age and experience. *Nature* **375**, 328-331.

Klein R., Parada L.F., Coulier F. and Barbacid M. (1989) trkB, a novel tyrosine protein kinase receptor expressed during mouse neural development. *EMBO J.* **8**, 3701-3709.

Klein R., Conway D., Parada L.F. and Barbacid M. (1990) The trkB tyrosine protein kinase gene codes for a second neurogenic receptor that lacks the catalytic kinase domain. *Cell* **61**, 647-656.

Klein R., Jing S., Nanduri V., O'Rourke E. and Barbacid M. (1991a) The trk proto-oncogene encodes a receptor for nerve growth factor. *Cell* **85**, 189-197

Klein R., Nanduri V., Jing S., Lamballe F., Tapley P., Bryant S., Cordon-Cardo C., Jones K.R., Reichardt L.F. and Barbacid M. (1991b) The trkB tyrosine protein kinase is a receptor for brain-derived neurotrophic factor and neurotrophin-3. *Cell* **66**, 395-403.

Klein R., Smeyne R.J., Wurst W., Long L.K., Auerbach B.A., Joyner A.L. and Barbacid M. (1993) Targeted disruption of the trkB neurotrophin receptor gene results in nervous system lesions and neonatal death. *Cell* **75**, 113-122.

Klein R., Silos-Santiago I., Smeyne R.J., Lira S.A., Brambilla R., Bryant S., Zhang L., Snider W.D. and Barbacid M. (1994) Disruption of the neurotrophin-3 receptor gene trkC eliminates Ia muscle afferents and results in abnormal movements. *Nature* **368**, 249-251.

Klein R. (1994) Role of neurotrophins in mouse neuronal development. *FASEB J.* **8**, 738-744.

Knipper M., Leung L.S., Zhao D. and Rylett R.J. (1994) Short-term modulation of glutamatergic synapses in adult rat hippocampus by NGF. *NeuroReport* **5**, 2433-2436.

Knüsel B., Beck K.D., Winslow J.W., Rosenthal A., Burton L.E., Widmer H.R., Nikolics K. and Hefti F. (1992) Brain-derived neurotrophic factor administration protects basal forebrain cholinergic but not nigral dopaminergic neurons from degenerative changes after axotomy in the adult rat brain. *J. Neurosci.* **12**, 4391-4402.

Knüsel B., Rabin S.J., Hefti F. and Kaplan D.R. (1994) Regulated neurotrophin receptor responsiveness during neuronal migration and early differentiation. *J. Neurosci.* **14**, 1542-1554.

Korsching S. (1993) The neurotrophic factor concept: a reexamination. *J. Neurosci.* **13**, 2739-2748.

Korte M., Carroll P., Wolf E., Brem G., Thoenen H. and Bonhoeffer T. (1995)

Hippocampal long-term potentiation is impaired in mice lacking brain-derived neurotrophic factor. *Proc. Natl. Acad. Sci. USA* **92**, 8856-8860.

Lamballe F., Klein R. and Barbacid M. (1991) trkC, a new member of the trk family of tyrosine protein kinases, is a receptor for neurotrophin-3. *Cell* **66**, 967-979.

Large T.H., Bodary S.C., Clegg D.O., Weskamp G., Otten U. and Reichardt L.F. (1986) Nerve Growth Factor gene expression in the developing rat brain. *Science* **234**, 352-355.

Lee K.-F., Huber L.J., Landis C.S., Sharpe A.H, Chao M.V. and Jaenisch R. (1992) Targeted mutation of the gene encoding the low affinity NGF receptor p75 leads to deficits in the peripheral sensory nervous system. *Cell* **69**, 737-749.

Lee K.-F., Bachman K., Landis S. and Jaenisch R. (1994) Dependence on p75 for innervation of some sympathetic targets. *Science* **263**, 1447-1449.

Leibrock J., Lottspeich F., Hohn A., Hofer M., Hengerer B., Masiakowski P., Thoenen H. and Barde Y.A. (1989) Molecular cloning and expression of brain-derived neurotrophic factor. *Nature* **341**, 149-152.

Lein E., Hohn A. and Shatz C.J. (1995) Reciprocal laminar localization and developmental regulation of BDNF and NT-3 mRNA during visual cortex development. *Soc. Neurosci. Abstr.* **21**, 706.17.

Leßmann V., Gottmann K. and Heumann R. (1994) BDNF and NT-4/5 enhance glutamatergic synaptic transmission in cultured hippocampal neurones. *NeuroReport* **6**, 21-25.

LeVay S., Stryker M.P. and Shatz C.J. (1978) Ocular dominance columns and their development in layer IV of the cat's visual cortex. *J. Comp. Neurol.* **179**, 223-244.

LeVay S., Wiesel T.N. and Hubel D.H. (1980) The development of ocular dominance columns in normal and visually deprived monkeys. *J. Comp. Neurol.* **191**, 1-51.

Levi A. and Alemà S. (1991) The mechanism of action of nerve growth factor. *Annu. Rev. Pharmacol. Toxicol.* **31**, 205-228.

Levi-Montalcini R. (1951) Selective growth-stimulating effects of mouse sarcomas on the sensory and sympathetic nervous system of chick embryo. *J. Exp. Zool.* **116**, 321-362.

Levi-Montalcini (1987) The Nerve Growth Factor: thirty-five years later. *EMBO J.* **6**, 2865-2867.

Li Y., Holtzman D.M., Kromer L.F., Kaplan D.R., Choua-Couzens J., Clary D.O., Knüsel B. and Mobley W.C. (1995) Regulation of TrkA and ChAT expression in developing rat basal forebrain: evidence that both exogenous and endogenous NGF regulate differentiation of cholinergic neurons. *J. Neurosci.* **15**, 2888-2905.

Lindholm D. (1994) Two promoters direct transcription of the mouse NT-3 gene. *Eur. J. Neurosci.* **6**, 1149-1159.

Lindsay R.M. (1988) The role of neurotrophic growth factors in development, maintenance and regeneration of sensory neurons. In: The making of the nervous system (Parnavelas J.G., Stern C.D., Stirling R.V., eds), pp 148-165. Oxford University Press, Oxford, UK.

Lindsay R.M., Thoenen H. and Barde Y.-A. (1985) Placode and neural crest-derived sensory neurons are responsive at early developmental stages to brain-derived neurotrophic factor. *Dev. Biol.* **112**, 319-328.

Lindsay R.M., Wiegand S.J., Altar C.A. and DiStefano P.S. (1994) Neurotrophic factors: from molecule to man. *Trends Neurosci.* **17**, 182-190.

Liu X., Ernfors P., Wu H. and Jaenisch R. (1995) Sensory but not motor neuron deficit in mice lacking NT4 and BDNF. *Nature* **375**, 238-241.

Loeb D.M., Tsao H., Cobb M.H. and Greene L.A. (1992) NGF and other growth factors induce an association betweeen ERK1 and the NGF receptor, gp140^{prototrk}. *Neuron* **9**, 1053-1065.

Lohof A.M., Ip N.Y. and Poo M. (1993)   Potentiating of developing neuromuscular synapses by the neurotrophins NT-3 and BDNF. *Nature* **363**, 350-353.

Lu B., Yokoama M., Dreyfus C.F. and Black I.B. (1991) Depolarizing stimuli regulate nerve growth factor gene expression in cultured hippocampal neurons. *Proc. Natl. Acad. Sci. USA* **88**, 6289-6292.

Maffei L., Berardi N., Domenici L., Parisi V. and Pizzorusso T. (1992) Nerve growth factor (NGF) prevents the shift in ocular dominance distribution of visual cortical neurons in monocularly deprived rats. *J. Neurosci.* **12**, 4651-4662.

Maisonpierre P.C., Belluscio L., Squinto S., Ip N.Y., Furth M.E., Lindsay R.M. and Yancopoulos G.D. (1990a) Neurotrophin-3: a neurotrophic factor related to NGF and BDNF. *Science* **247**, 1446-1451.

Maisonpierre P.C., Belluscio L., Friedman R.F., Alderson S.J., Wiegand M.E., Furth R.M., Lindsay R.M. and Yancopoulos G.D. (1990b) NT-3, BDNF and NGF in the developing rat nervous system: parallel as well as reciprocal patterns of expression. *Neuron* **5**, 501-509.

Martin-Zanca D., Hughes S.H. and Barbacid M. (1986) A human oncogene formed by the fusion of truncated tropomyosin and protein tyrosine kinase sequences. *Nature* **319**, 743-748.

Martin-Zanca D., Oskam R., Mitra G., Copeland T. and Barbacid M. (1989) Molecular and biochemical characterization of the human *trk* proto-oncogene. *Mol. Cell. Biol.* **9**, 24-33.

Martin-Zanca D., Barbacid M. and Parada L.F. (1990) Expression of the trk proto-oncogene is restricted to the sensory cranial and spinal ganglia of neural crest origin in mouse development. *Genes Dev.* **4**, 683-694.

McAllister A.K., Lo D.C. and Katz L.C (1995) Neurotrophins regulate dendritic growth in developing visual cortex. *Neuron* **15**, 791-803.

McCormick F. (1994) Activators and effectors of the *ras* p21 protein. *Curr. Opin. Genet. Dev.* **4**, 71-76.

McDonald N.Q., Lapatto R., Murray-Rust J., Gunning J., Wlodawer A. and Blundell T.L. (1991) New protein fold revealed by 2.3 Å resolution crystall structure of nerve growth factor. *Nature* **354**, 411-414.

Meakin S.O. and Shooter E.M. (1992) The nerve growth factor family of receptors. *Trends Neurosci.* **15**, 323-331.

Merlio J.P., Ernfors P., Jaber M. and Persson H. (1992) Molecular cloning of rat trkC and distribution of cells expressing messenger RNAs for members of the trk family in the rat central nervous system. *Neuroscience* **51**, 513-532.

Metsis M., Timmusk T., Arenas E. and Persson H. (1993) Differential usage of multiple brain-derived neurotrophic factor promoters in the rat brain following neuronal activation. *Proc. Natl. Acad. Sci. USA* **90**, 8802-8806.

Middlemas D.S., Lindberg R.A. and Hunter T. (1991) trkB, a neural receptor protein-tyrosine kynase: evidence for a full-length and two truncated receptors. *Mol. Cell. Biol.* **11**, 143-153.

Middlemas D.S., Meisenhelder J. and Hunter T. (1994) Identification of TrkB autophosphorylation sites and evidence that phospholipase C-g1 is a substrate of the TrkB receptor. *J. Biol. Chem.* **269**, 5458-5466.

Miranda R.C., Sohrabji F. and Toran-Allerand C.D. (1993) Neuronal colocalization of mRNAs for neurotrophin and their receptors in the developing central nervous system suggests a potential for autocrine interactions. *Proc. Natl. Acad. Sci. USA* **90**, 6439-6443.

Morgan J.L. and Curran T. (1991) Stimulus-transcription coupling in the nervous system: involvement of the inducible proto-oncogenes fos and jun. *Annu. Rev. Neurosci.* **14**, 421-451.

Nawa H., Carnahan J. and Gall C. (1995) BDNF protein measured by a novel enzyme immunoassay in normal brain and after seizure: partial disagreement with mRNA levels. *Eur. J. Neurosci.* **7**, 1527-1535.

Nedivi E., Fieldust S., Theill L.E and Hevroni D. (1996) A set of genes expressed in response to light in the adult cerebral cortex and regulated during development. *Proc. Natl. Acad. Sci. USA* **93**, 2048-2053.

Neeper S.A., Gómez-Pinilla F., Choi J. and Cotman C. (1995) Exercise and brain neurotrophins. *Nature* **373**, 109.

Neve R.L. and Bear M.F. (1989) Visual experience regulates gene expression in the developing striate cortex. *Proc. Natl. Acad. Sci. USA* **86**, 4781-4784.

Obermeier A., Bradshaw R.A., Seedorf K., Choidas A., Schlessinger J. and Ullrich A. (1994) Neuronal differentiation signals are controlled by nerve growth factor receptor/trk binding sites for SHC and PLCγ. *EMBO J.* **13**, 1585-1590.

Oppenheim R.W. (1991) Cell death during development of the nervous system. *Ann Rev. Neurosci.* **14**, 453-501.

Parsadian A., Lindholm D., Thoenen H. and Riekkinen P.J. (1993) The induction of LTP increases BDNF and NGF mRNA but decreases NT-3 mRNA in dentate gyrus. *NeuroReport* **4**, 895-898.

Patterson S.L., Grover L.M., Schwartzkroin P.A. and Bothwell M. (1992) Neurotrophin expression in rat hipocampal slices: a stimulus paradigm inducing LTP in CA1 evokes increases in BDNF and NT-3 mRNAs. *Neuron* **9**, 1081-

1088.

Paxinos G. and Watson C. (1986) *The rat brain in stereotaxic coordinates*, 2nd ed. Academic Press, Sydney.

Peterson C., Venero J.L., Kaplan D.R. and Knüsel B. (1995) Does NGF remain biologically active in rat brain tissue for several days? *Soc. Neurosci. Abstr.* **21**, 224.9.

Phillips H.S., Hains J.M., Laramee G.R., Rosenthal A. and Winslow J.W. (1990) Widespread expression of BDNF but not NT3 by target areas of basal forebrain cholinergic neurons. *Science* **250**, 290-294.

Pioro E.P. and CuelloA.C. (1990) Distribution of nerve growth factor receptor-like immunoreactivity in the adult rat central nervous system. Effect of colchicine and correlation with the cholinergic system - I. Forebrain. *Neuroscience* **34**, 57-87.

Pizzorusso T., Fagiolini M., Fabris M., Ferrari G. and Maffei L. (1994) Schwann cells transplanted in the lateral ventricles prevent the functional and anatomical effects of monocular deprivation in the rat. *Proc. Natl. Acad. Sci. USA* **91**, 2572-2576.

Pizzorusso T., Porciatti V., Tseng J., Aebischer P. and Maffei L. (1995) NGF replaces visual experience in the development of the rat visual cortex. *Soc. Neurosci. Abstr.* **21**, 119.11.

Puma P., Buxser S.E., Watson L., Kelleher D.J. and Johnson G.L. (1983) Purification of the receptor for nerve growth factor from A875 melanoma cells by affinity chromatography. *J. Biol. Chem.* **258**, 3370-3375.

Purves D. (1986) The trophic theory of neural connections. *Trends Neurosci.* **9**, 486-489.

Radeke M.J., Misko T.P., Herzenberg L.A. and Shooter E.M. (1987) Gene transfer and molecular cloning of the rat nerve growth factor receptor. *Nature*

325, 593-597.

Raffioni S., Bradshaw R.A. and Buxser S.E. (1993) The receptors for nerve growth factor and other neurotrophins. *Ann. Rev. Biochem.* **62**, 823-850.

Rakic P. (1976) Prenatal genesis of connections subserving ocular dominance in the rhesus monkey. *Nature* **261**, 467-471.

Reese B.E. (1988) "Hidden lamination" in the dorsal lateral geniculate nucleus: the functional organization of this thalamic region in the rat. *Brain Res. Rev.* **13**, 119-137.

Reiter H.O. and Striker M.P. (1988) Neural plasticity without postsynaptic action potentials: less-active inputs become dominant when kitten visual cortex cells are pharmacologically inhibited. *Proc. Natl. Acad. Sci. USA* **85**, 3623-3627.

Riddle D.R., Lo D.C. and Katz L.C. (1995) NT-4-mediated rescue of lateral geniculate neurons effects of monocular deprivation. *Nature* **378**, 189-191.

Ringstedt T., Lagercrantz H. and Persson H. (1993) Expression of members of the trk family in the developing postnatal rat brain. *Dev. Brain Res.* **72**, 119-131.

Roback J.D., Marsh N., Downen M., Palfrey C.H. and Wainer B.H. (1995) BDNF-activated signal transduction in rat cortical glial cells. *Eur. J. Neurosci.* **7**, 849-862.

Rodriguez-Tébar A., Jeffrey P.L., Thoenen H. and Barde Y.A. (1989) The survival of chick retinal ganglion cells in response to brain-derived neurotrophic factor depends on their embryonic age. *Dev. Biol.* **136**, 296-303.

Rodriguez-Tébar A., Dechant G. and Barde Y.A. (1990) Binding of brain-derived neurotrophic factor to the nerve growth factor receptor. *Neuron* **4**, 487-492.

Rodriguez-Tébar A., Dechant G., Gotz R. and Barde Y.A. (1992) Binding of neurotrophin-3 to its neuronal receptors and interactions with nerve growth factor and brain-derived neurotrophic factor. *EMBO J.* **11**, 917-922.

Rosenthal A., Goeddel D.V., Nguyen T., Lewis M., Shih A., Laramee G.R., Nikolics K. and Winslow J.W (1990) Primary structure and biological activity of a novel human neurotrophic factor. *Neuron* **4**, 767-773.

Rossi F.M., Bozzi Y., Pizzorusso T., Maffei L. and Yan Q. (1996) Brain-derived neurotrophic factor (BDNF) immunoreactivity in the rat visual cortex. *Keyston Symposium on Signaling in neuronal development Abs.*, 138.

Sambrook J., Fritsch E.F. and Maniatis T. (1989) Molecular cloning - A laboratory manual. Cold Spring Harbor Laboratory Press, 2nd ed.

Schechter A.L. and Bothwell M.A. (1981) Nerve growth factor receptors on PC12 cells: evidence for two receptor classes with different cytoskeletal association. *Cell* **24**, 867-874.

Schlessinger J. (1994) SH2/SH3 signaling proteins. *Curr. Opin. Gene. Dev.* **4**, 25-30.

Schlessinger J. and Ullrich A. (1992) Growth factor signaling by receptor tyrosine kinases. *Neuron* **9**, 383-391.

Schoups A.A., Elliott R.C., Friedman W.J. and Black I.B. (1995) NGF and BDNF are differentially modulated by visual experience in the developing geniculocortical pathway. *Dev. Brain Res.* **86**, 326-334.

Schuman E.M. and Madison D.V. (1994) Nitric oxide and synaptic function. *Annu. Rev. Neurosci.* **17**, 153-183.

Scott J., Selby M., Urdea M., Quiroga M., Bell G.I. and Rutter W.J. (1983) Isolation and nucleotide sequence of a cDNA encoding the precursor of mouse nerve growth factor. *Nature* **302**, 538-540.

Sefton A.J. and Dreher B. (1985) Visual system. In: *The rat nervous system* (ed.

G. Paxinos) pp. 169-221. Academic Press, Sydney.

Segal R.A. and Greenberg M.E. (1996) Intracellular signaling pathways activated by neurotrophic factors. *Annu. Rev. Neurosci.* **19**, 463-489.

Seroogy K.B., Lundgren K.H., Tran T.M., Guthrie K.M., Isackson P.J. and Gall C.M. (1994) Dopaminergic neurons in rat ventral midbrain express brain-derived neurotrophic factor and neurotrophin-3 mRNAs. *J. Comp. Neurol.* **342**, 321-334.

Shatz C.J. (1990) Impulse activity and the patterning of connections during CNS development. *Neuron* **5**, 745-756.

Shatz C.J. and Stryker M.P. (1978) Ocular dominance in layer IV of the cat's visual cortex and the effects of monocular deprivation. *J. Physiol. (London)* **281**, 267-283.

Sheng M. and Greenberg M.E. (1990) The regulation and function of c-fos and other immediate early genes in the nervous system. *Neuron* **4**, 477-485.

Sheng M., McFadden G. and Greenberg M.E. (1990) Membrane depolarization and calcium induce c-fos transcription via phosphorylation of transcription factor CREB. *Neuron* **4**, 571-582.

Sherman S.M. and Spear P.D. (1982) Organization of visual pathways in normal and visually deprived cats. *Physiol. Rev.* **62**, 738-855.

Smeyne R.J., Klein R., Schnapp R., Long L.K., Bryant S., Lewin A., Lira S.A. and Barbacid M. (1994) Severe sensory and sympathetic neuropathies in mice carrying a disrupted trk/NGF receptor gene. *Nature* **368**, 246-249.

Snider W.D. (1994) Functions of the neurotrophins during nervous system development: what the knockouts are teaching us. *Cell* **77**, 627-638.

Sobreviela T., Clary D.O., Reichardt L.F., Brandabur M.M., Kordower J.H. and

Mufson E.J. (1994) TrkA-immunoreactive profile in the central nervous system: colocalization with neurons containing p75 nerve growth factor receptor, choline acetyltransferase, and serotonine. *J. Comp. Neurol.* **350**, 587-611.

Squinto S.P., Stitt T.N., Aldrich T.H., Davis S., Bianco S.M., Radziejewski C., Glass D.J., Masiakowski P., Furth M.E., Valenzuela D.M., DiStefano P.S. and Yancopoulos G.D. (1991). trkB encodes a functional receptor for brain-derived neurotrophic factor and neurotrophin-3 but not nerve growth factor. *Cell* **65**, 885-893.

Sretavan D.W. and Shatz C.J. (1986) Prenatal development of retinal ganglion cell axons: segregation into eye-specific layers. *J. Neurosci.* **6**, 234-251.

Steininger T.L., Wainer B.H., Klein R., Barbacid M. and Palfrey H.C. (1993) High-affinity nerve growth factor receptor (Trk) immunoreactivity is localized in cholinergic neurons of the basal forebrain and striatum of the adult rat brain. *Brain Res.* **612**, 330-335.

Stephens R.M., Loeb D.M., Copeland T.D., Pawson T., Greene L.A. and Kaplan D.R. (1994) Trk receptors use redundant signal transduction pathways involving SHC and PLC-γ1 to mediate NGF responses. *Neuron* **12**, 691-705.

Stryker M.P. and Harris W. (1986) Binocular impulse blockade prevents the formation of ocular dominance columns in cat visual cortex. *J. Neurosci.* **6**, 2117-2133.

Stryker M.P. and Strickland S.L. (1984) Physiological segregation of ocular dominance columns depends on the pattern of afferent electrical activity. *Invest. Ophtalmol. Vis. Sci. (suppl.)* **25**, 278.

Südhof T. (1995) The synaptic vesicle cycle: a cascade of protein-protein interactions. *Nature* **375**, 645-653.

Sutter A., Riopelle R.J., Harris-Warrick R.M. and Shooter E.M. (1979) Nerve growth factor receptors. Characterization of two distinct classes of binding sites on chick embryo sensory ganglia cells. *J. Biol. Chem.* **254**, 5972-5982.

Thoenen H. (1995) Neurotrophins and neuronal plasticity. *Science* **270**, 593-598.

Thoenen H, Bandtlow C. and Heumann R. (1987) The physiological function of Nerve Growth factor in the central nervous system: comparison with the periphery. *Rev. Physiol. Biochem Pharmacol.* **109**, 145-178.

Timmusk T., Belluardo N., Metsis M. and Persson H. (1993a) Widespread and developmentally regulated expression of neurotrophin-4 mRNA in rat brain and peripheral tissues. *Eur.J. Neurosci.* **5**, 605-613.

Timmusk T., Palm K., Metsis M., Reintam T., Paalme V., Saarma M. and Persson H. (1993b) Multiple promoters direct tissue-specific expression of the rat BDNF gene. *Neuron* **10**, 475-489.

Timmusk T., Lendahl U., Funakoshi H., Arenas E., Persson H. and Metsis M. (1995) Identification of brain-derived neurotrophic factor promoter regions mediating tissue-specific, axotomy-, and neuronal activity-induced expression in transgenic mice. *J. Cell. Biol.* **128**, 185-199.

Toledo-Aral J.J., Brehm P., Halegoua S. and Mandel G. (1995) A single pulse of nerve growth factor triggers long-term neuronal excitability through sodium channel gene induction. *Neuron* **14**, 607-611.

Ullrich A., Gray A., Berman C. and Dull T.J. (1983) Human β-nerve growth factor gene sequence highly homologous to that of mouse. *Nature* **303**, 821-825.

Valenzuela D.M., Maisonpierre P.C., Glass D.J., Rojas E., Nuñez L., Kong Y., Gies D.R., Stitt T.N., Ip N.Y. and Yancopoulos G.D. (1993) Alternative forms of rat trkC with different functional capabilities. *Neuron* **10**, 963-974.

Wetmore C., Ernfors P., Persson H. and Olson L. (1990) Localization of brain-derived neurotrophic factor mRNA to neurons in the brain by *in situ* hybridization.

Exp. Neurol. **109**, 141-152.

Wetmore C., Olson L. and Bean A.J. (1994) Regulation of Brain-derived neurotrophic factor (BDNF) expression and release from hippocampal neurons is mediated by non-NMDA type glutamate receptors. *J. Neurosci.* **14**, 1688-1700.

Wiesel T.N. and Hubel D.H. (1963a) Effects of visual deprivation on morphology and physiology of cells in the cat's lateral geniculate nucleus. *J. Neurophysiol.* **26**, 979-993.

Wiesel T.N. and Hubel D.H. (1963b) Single cell responses in striate cortex of kittens deprived of vision in one eye. *J. Neurophysiol.* **26**, 1003-1017.

Wiesel T.N. and Hubel D.H. (1965) Comparison of the effects of unilateral and bilateral eye closure on cortical unit responses in kittens. *J. Neurophysiol.* **28**, 1029-1040.

Wilkinson D.G. and Green J. (1990) *In situ* hybridization and three-dimensional reconstruction of serial sections. In: *Practical Approach Series: postimplantation Mouse Embryos: A Practical Approach* (D. Rickwood and D. Cockcroft, eds.), pp. 155-171. Oxford: IRL Press.

Worley P.F., Christy B.A., Nakabeppu Y., Bhat R.V., Cole A.J. and Baraban J.M. (1991) Constitutive expression of *zif268* in neocortex is regulated by synaptic activity. *Proc. Natl. Acad. Sci. USA* **88**, 5106-5110.

Zafra F., Hengerer B., Leibrock J., Thoenen H. and Lindholm D. (1990) Activity dependent regulation of BDNF and NGF mRNAs in the rat hippocampus is mediated by non-NMDA glutamate receptors. *EMBO J* . **9**, 3545-3550.

Zafra F., Castrén E., Thoenen H. and Lindholm D. (1991) Interplay between glutamate and γ-aminobutyric acid transmitter systems in the physiological regulation of brain-derived neurotrophic factor and nerve growth factor synthesis in hippocampal neurons. *Proc Natl Acad Sci USA* **88,** 10037-10041.

Zafra F., Lindholm D., Castrén E., Hartikka J. and Thoenen H. (1992) Regulation of brain-derived neurotrophic factor and nerve growth factor mRNA in primary cultures of hippocampal neurons and astrocytes. *J. Neurosci.* **12**, 4793-4799.

Zanellato A., Comelli M.C., Dal Toso R. and Carmignoto G. (1993) Developing rat retinal ganglion cells express the functional NGF receptor p140trkA. *Dev. Biol.* **159**, 105-113.

Zilles K., Wree A., Schleicher A. and Divac I. (1984) The monocular and binocular subfields of the rat's primary visual cortex: a quantitative morphological approach. *J. Comp. Neurol.* **226**, 391-402.

ACKNOWLEDGEMENTS

My four years at Scuola Normale Superiore have been a very stimulating experience and I am therefore very grateful to all the people I have met in that period.

I wish to express my thanks to Professor Lamberto Maffei and Professor Giuseppina Barsacchi, who have given me the chance of working in a field of extreme interest and actuality and have followed my work with all their competence. I also would like to thank Professor Dan Lindholm (Max Planck Institute für Psychiatrie, Martinsried, Germany) for his helpful and critical comments on the preliminary version of my Thesis.

Dr. Delio Mercanti helped me to set up some experimental protocols when I spent some weeks at the Istituto di Neurobiologia del CNR in Rome; Dr. Maria Cristina Comelli and Dr. Alessandro Negro (FIDIA Research Laboratories, Padova, Italy), Dr. Roberta Possenti (Istituto di Neurobiologia del CNR, Rome, Italy), Dr. Qiao Yan (Amgen Corp., USA), Dr Eero Castrén (Max Planck Institute für Psychiatrie, Martinsried, Germany) and Genentech Laboratories (San Francisco, CA, USA) kindly granted the use of experimental reagents. Mr. Alberto Bertini and Piero Taccini (Istituto di Neurofisiologia del CNR, Pisa) gave an excellent assistance in photographic work.

I am grateful to all my friends at the BMCF (Laboratorio di Biologia Cellulare e dello Sviluppo, University of Pisa) and at the Istituto di Neurofisiologia del CNR in Pisa, with whom I shared rewarding experiences both in the laboratory and outside: in particular, I would like to thank Dr. Federico Cremisi, Dr. Tommaso Pizzorusso and Dr. Francesco Mattia Rossi for their constant help; Dr. Paolo Malatesta gave invaluable advice as to the computer graphs and drawings of the Thesis.

More than anyone else, Dr. Simona Casarosa has constantly been at my side and has given me day by day aid and encouragement.

Elenco delle Tesi di perfezionamento della Classe di Scienze
pubblicate dall'Anno Accademico 1992/93

HISAO FUJITA YASHIMA, *Equations de Navier–Stokes stochastiques non homogènes et applications*, 1992.

GIORGIO GAMBERINI, *The Minimal supersymmetric standard model and its phenomenological implications*, 1993.

CHIARA DE FABRITIIS, *Actions of Holomorphic Maps on Spaces of Holomorphic Functions*, 1994.

CARLO PETRONIO, *Standard Spines and 3-Manifolds*, 1995.

MARCO MANETTI, *Degenerations of Algebraic Surfaces and Applications to Moduli Problems*, 1995.

ILARIA DAMIANI, *Untwisted Affine Quantum Algebras: the Highest Coefficient of $detH_\eta$ and the Center at Odd Roots of 1*, 1995.

FABRIZIO CEI, *Search for Neutrinos from Stellar Gravitational Collapse with the MACRO Experiment at Gran Sasso*, 1995.

ALEXANDRE SHLAPUNOV, *Green's Integrals and Their Applications to Elliptic Systems*, 1996.

ROBERTO TAURASO, *Periodic Points for Expanding Maps and for Their Extensions*, 1996.

YURI BOZZI, *A study on the activity-dependent expression of neurotrophic factors in the rat visual system*, 1997.

MARIA LUISA CHIOFALO, *Screening effects in bipolaron theory and high-temperature superconductivity*, 1997.

Pantograf s.n.c. - Via alla Stazione di Voltri, 6 R. - Genova
Finito di stampare nel gennaio 1997